**THEORY OF EVERYTHING IN PHYSICS AND THE UNIVERSE**

1st edition published in Australia by Valentin Malinov 2020
2nd edition published in Australia by Valentin Malinov 2021

Copyright © Valentin Malinov 2021
All Rights Reserved

ISBN: 978-0-6480127-6-4 (pbk)
ISBN: 978-0-6480127-7-1 (ebk)

Details available from The National Library of Australia

Artwork and photography by Valentin Malinov © 2021

Published with the assistance of Publicious Book Publishing
www.publicious.com.au

# Contents

## INTRODUCTION

We, the modern generation are proud of the achievements of modern science and we honestly believe that our science is providing us with a very good explanation of the World. Many scientists are stating that in the last one hundred years we have achieved more than humanity has achieved in all its existence.

I agree that the accumulation of scientific data in this short period is enormous, but I seriously doubt that we are using correctly this data and we understanding our World better than our ancestors. For example, the ancient Greek philosophers without any scientific instruments were able to figure out the correct structure of the Solar system, the correct size of the visible planets, and their distances. They were able to understand the chemical reactions and the correct structure of matter - down to its atomic structure. These colossi of human intelligence have been able to define the fundamental principles of the World. They understood the necessity of Democracy and Ethics in civilized societies to function. - These are the vital principles, which we stubbornly refuse to understand and implement in our life. It is sad to make such a statement, but I have no other choice - Currently, even with our sophisticated instruments and enormous scientific data, we do not have a better understanding of the World than our ancestors. In some aspects even we are trailing them because most of our concepts for the physical order of the Universe are incorrect. - We don't know what is Energy, we don't know what is Electromagnetism, we don't know the principle of Physical Attraction, we don't know what Space is, we don't know what Time is, we don't know where the Gravity is coming from, we don't know what is Strong Nuclear Force, and even we have an incorrect model of the atomic nucleus. This lack of fundamental knowledge is a result of the incorrect fundamental principles adopted by the dogma of the "Standard Model of Physics." It is obvious, that when our fundamental principles are incorrect nothing works and we cannot construct a correct picture of the World and we cannot understand it! - We have become masters of self-deception and mass media is proud of this achievement.

I will try to put end to this uncertainty and will explain all these currently unknown fundamental elements of the Physical organization of the Universe in simple and easy-to-understand language.

I hope, to make a step forward in our understanding of the World, and this understanding brings us a desire to understand who we are, where we come from, and what is our role, purpose, and future in the Universe.

A few years ago I was able to define what a single Dimension is. This insignificant on first look discovery turn out to be the most significant discovery I can imagine. The understanding of the nature of a single dimension makes me able to understand what really Time is because Time also is a single dimension. Then, when I was armed with the knowledge of what Time is and having a correct understanding of what is a single dimension, it becomes very easy to figure out how Space is formed, how many dimensions are involved, and how Time providing the irreversibility of the Physical processes in the Universe. Then, the knowledge of the space structure has reviled where all the matter of the Universe is coming from. I was shocked and amassed of how one insignificant on first look discovery brings an avalanche of a new understanding of the fundamental structures of the World. The secrets of the Universe start falling in place one by one and in a short time everything falls in its place and I had and enjoying the new and complete picture of our World full of logic, sense, and beauty.

I am not claiming that I am the smartest person in the World; - I am very far from such an assumption. I am just trying to explain how I was able to define all the fundamental elements of the Universe - something which the army of well-paid scientists provided with the best scientific instruments wasn't able to do.

I manage to succeed because I am not afraid to trust logic and facts, even when they contradicting the well-established view.

The real break true came when I was able to define correctly the first two fundamental elements - Single Dimension and Time. These two elements were giving me a good starting point and when I started with two correctly defined fundamental elements, then everything else gradually and easily has fallen in its place. - This is how I was able to produce the most elusive and wanted theory in Physics and Astronomy - the so-called "Theory of Everything" - (TOE). The short version of the Theory had been published last year (2019) in a leading Russian scientific magazine. A small part of it you can find in my book "Myths Lies Illusions and the Way Out."

In the published articles I didn't include the subjects of Strong Nuclear Force, because I was afraid that if I revile more details, this will lead to an understanding of Gravity. We know that most of the new discoveries usually first are used for military purposes. I didn't want to revile the origin of Gravity, because I was afraid that Gravity could become a tool of mass murder and destruction in the hands of the military machines of the greatest Countries. After careful study and consideration, I conclude that to be able to achieve practical use of Gravity is necessary about 50 years of scientific research and experiments. I could make a load of money if I have sold the secret of Gravity. There are enough customers for such a valuable secret, but I don't care! I don't want their money; I just want to make my free contribution to the knowledge and prosperity of humanity.

Now, when I am not afraid for your safety my friends, I will revile the rest of the secrets of the physical organization of our Universe. - (Val)

## THEORY OF EVERYTHING IN PHYSICS AND THE UNIVERSE

In tribute of the greatest man in human history – Giordano Bruno, who's understanding of the Universe still is too advanced even for the present generation.

# Where the problems are coming from

The 'Standard Model' of astrophysics reluctantly has departed from the practice of using correct scientific facts and methods in pursuit of providing a scientific explanation of our World. The leading theories are full of bizarre assumptions, which violate the laws of Physics, observational data, facts, and healthy logic.

The scientific instruments are getting more and more sophisticated and are providing us with a rich amount of data and observational evidence for the structure of our world. No matter how accurate is the collected data, if this data is not considered correctly with the Law of Physics and with healthy logic, the avalanche of new information will create more confusion rather than explain anything. - Ultimately this exactly is what had happened! The efforts of the scientific elite to keep the old theories in place lead to ignorance of the law of Physics and a range of important scientific facts and observations for building a correct picture of our World!

Many scientists are stating that we have a crisis in science, but not one of them dares to identify what it is. So... why are they talking about crisis and never reveal where the actual problem is? Where this problem is coming from and what we should do to correct it?

I will start the explanation of the crisis with a few well-known facts:

One hundred years ago, when the fundamental theories for the structure of the Universe have been established, the scientists had no idea, that the visible part of the Universe is less than 1% of its real matter content.

Later, with the development of radio-astronomy, we have discovered, that the diffused matter, dust, and plasma in the interstellar medium represent at least 99% of the matter content of the Universe. - (In this matter content of the Universe is not included the hypothetical Dark Matter and Dark Force). – See the image below.

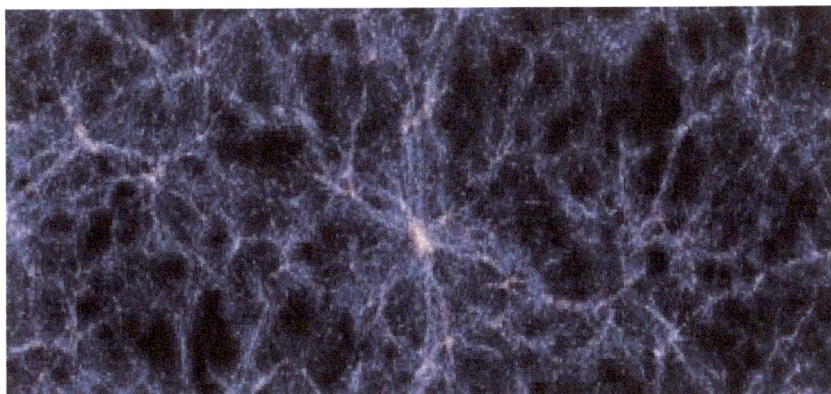

This is how the "Empty" space looks through the eye of Chandra space telescope. Why do we continue ignoring the existence of the "invisible" to the eye interstellar filaments?

From the beginning of the 20<sup>th</sup> century until now, the foundation of astrophysics still is based only on this 1% of the visible matter content and is not taken into account the superior electromagnetic force, which is ruling the invisible 99% universal matter and is influencing to great extent the energy exchange and movement of the visible part of the Universe. The electromagnetic force is many billion times stronger than gravity, and if we would like to have a realistic model and understanding of the Universe, we should include in our model all known to us matter, forces and components. Unfortunately, the current situation in physics is similar to a scenario, where we are given a microscope, to find the sense and beauty of a fully covered Rembrandt painting. Currently, we are studying the Universe with the biggest microscope on Earth – the Large Hadrons Collider. We are smashing particles with the hope to find and explain the picture and dynamics of the colossal universal structures without taking into account the 99% of matter, the dominant force of electromagnetism, and its dynamics! As a result of this approach, we are bombarded with bizarre theories for non-existing phenomena's as Gluons, Gravitons, Virtual Particles, Mass-less Particles, Dark Matter, Dark force, Higgs boson, Universe from Nothing,  50% Missing Matter, Expanding Universe, Neutron stars, Black holes, Dark light and mysteriously "self-controlled" Nuclear Fusion of the Sun and Stars.

We are spending enormous manpower and resources needlessly to keep in place old and impossible theories, which have nothing to do with the reality of

the World! I just would like to mention, that all those puzzles are created because the "Standard Model" of Physics is based on the wrong concept. To back up this wrong concept, academic science is forced to ignore the Law of Physics, to ignore and manipulate facts, observations, and results.
In this book, such ways of selective considerations and ignorance of the scientific facts will be avoided. The careful consideration and logical analysis of all available facts are leading to completely different results, different conclusions, and a completely different picture of our amazing Universe.


## THE SIX FUNDAMENTAL ELEMENTS OF THE UNIVERSE

The Universe is constructed of three material and three non-material components:
Space, Time, and Matter are the physical (or the material) components of the Universe.
Consciousness, Universal (Quantum) Information, and the Law of Physics are non-material components of the Universe.
The politically adapted materialistic philosophy is dictating to the scientific community to consider only the material aspect of the Universe.
On the exact opposite, the religious community is considering mostly the non-material aspect of the Universe.
Consideration of only half of the building blocks of the Universe has lead to Materialism and Spiritualism.
Both philosophies are a primitive and incomplete way to deal with and explain the properties of the Universe! We will avoid such a selective approach and will not be afraid to consider all fundamental elements in our quest for the correct understanding of the Universe.

- THE LAW OF PHYSICS – The Laws of Physics are a sophisticated informational order which can be a product only of an intelligent creative mind, or logic possessing phenomenon. The laws of Physics precede the existence of the material components of the Universe because they cannot exist without physical order. This logical conclusion is revealing the consecutive design pattern of the fundamental structures of the Universe.
The Law of Physics is the "Software of the Universe" - It is the intellectual

product of the Universal Consciousness. The Law of Physics is acting as a designed plan and functional order of The Universe!

The carefully inserted safeguard mechanisms in the physical order of the Universe (explained in my previous book) are not allowing us to know and control, change or manipulate the fundamental properties of the Universe. - The law of physics is one of them. The embedded limit of knowledge is not allowing us to know where the Law of Physics is situated and how it is processing the quantum information and how is controlling the physical processes of the Universe. The limit of knowledge is not allowing us to change or manipulate the Law of Physics with very good reason.

- QUANTUM INFORMATION – for simplicity I am using this term to describe the Universal Informational link between all parts of matter and structures of the Universe. Quantum information is not a physical phenomenon!  For this reason, the physical processes of the Universe cannot affect the information! The law of Physics states that you cannot create, destroy, copy, or duplicate quantum information! Quantum information is the informational link between each particle and each physical structure of The Universe! The Quantum Information travels instantly at any distance and is not affected by Time, because Time is an energy-related substance and belongs to the physical part of the Universe. Quantum information is protected with an incredibly sophisticated Unbreakable Code which is one of the embedded "Limits of Knowledge". The information is the invisible tool or is the "hand" of the Law of Physics and is the mechanism to keep the order of The Universe! The quantum information is stored in the Universal Consciousness! The information does not belong and is not stored in the matter – (How is the official scientific view).  This fact explains why and how the particles are able to "remember" their original property and spin when they change their physical forms!

- CONSCIOUSNESS – The Universal Consciousness is not a material substance and is not affected by the physical processes of the Universe. Consciousness is a "symbioses" of intelligent information and unknown to us a form of energy. The origin and structure of consciousness currently are unknown, but its presence is everywhere, it is infiltrated into every particle, and every living cell! You cannot create, destroy, duplicate, or manipulate Consciousness! Consciousness is the first fundamental phenomenon of the Universe and is responsible for its existence and the existence of all other components of the Universe. Consciousness is the origin of the Laws of Physics. It is the memory storage and origin of the Quantum Information, which is the actual tool or is the physical mechanism of the Laws of Physics to be implemented into every part and every particle of the Universe.

In association with our computers, Consciousness is correspondent to the hard drive, and the Law of Physics is the "Window Software". Quantum information is a product of those two components we just mentioned. Consciousness is not a product of our mind! The fact that the plants and all living organisms possess consciousness, and most of them have no brain, or nervous system is the actual proof that our consciousness also is not a product of our brain! Our mind is just our processor, which is using consciousness as every other organism and every part of the Universe! The capacity of the "processor" determines the quantity of consciousness it can possess and use.

- TIME – Is a single dimension – it is a vector; it is dynamic progressive direction without volume, without beginning or end. We can characterize Time also as an active propagating (uniform) energy field.

- SPACE – Our space is a six-dimensional physical phenomenon, (or component) of The Universe! - (The explanation of space is below).
As everything possesses physical property in The Universe, Space also is made by the common fundamental building substance – energy. The three physical components of The Universe are Space, Time, and Matter. These fundamental physical components share a common building element, which is energy. Energy and information are the unifying elements of the material structure of The Universe!

# WHAT IS A SINGLE DIMENSION?

I will start the explanation with consideration of our current understanding of spatial dimensions and particularly with our understanding of what a single space dimension is. There is a range of conflicting statements about the nature and property of a single spatial dimension as "Higher dimensions", "Inductive dimensions", "Tangled dimensions", "Minkowski dimension", Hausdorff dimension" and ... an endless range of scientific speculations. It is obvious that our current understanding of dimensions is based on conflicting ideas and is foggy and chaotic.

Fortunately, we have a very good example and a relatively good understanding of the nature of a single dimension – which is Time!

Time is described as an arrow, (or vector), which has only one direction - from (here and now) toward the future. This is a well-known fact, but nobody is taking real notice of this. If we apply this knowledge to the understanding of what exactly a single spatial Dimension is, it won't be difficult to realize that Single Spatial Dimension is a Vector, it is <u>dynamic Direction</u> - It is a progressive directionally oriented phenomenon. The assumption of the creators of 'String Theory' that their "Strings" are one-dimensional is a clear indication that modern science doesn't understand what a single dimension is! - Their "Strings" suppose to be material objects. - We know that any material object will have volume and will be three-dimensional no matter how small it is! On top of this, we can say that the "strings" have a specific length and two ends. - Any object with two ends can be measured in two opposite directions, but a <u>single dimension cannot be measured in reverse, because is one-directional!</u> - Any physical object or material substance also cannot be two-dimensional, because matter inevitably has a volume! A two-dimensional configuration has only a surface without volume and also cannot accommodate physical objects! Two-dimensional configuration is a hypothetical scenario because we never observing two-dimensional structures in our everyday life.

A single dimension is a <u>one-directional vector starting from the reference point out</u>. I am including the reference point for a very good reason, which we will consider later. The reference point is playing a significant role in defying and understanding Space, Time, and Matter. The significance of the reference point is that everything in the Universe – every particle or atom must have its own strictly defined specific position in space, other ways will be chaos!

A single dimension is a direction only! The single dimension is a vector! I is an arrow and is direction without volume! A single dimension has no point of beginning or point of end! It is a direction without volume! When we represent the well-known property of Time as a vector, we will realize that in the Time dimension the term "Now" is the reference point from which Time is propagating forward - (toward the future). <u>That means that a single dimension is an arrow - it is a direction that starting from the reference point out and cannot be measured in reverse</u>! - This is the simplest and the greatest realization, which has opened the door for the correct understanding of the entire Universe!

## HOW MANY DIMENSIONS ARE FORMING OUR SPACE?

Modern science insisting that our space is three-dimensional, but no evidence is provided in support of this scenario.
To see how credible is this scenario let's apply our knowledge of what a single dimension is to a three-dimensional configuration to see what will happen?
When we start applying our knowledge of what a single dimension is to the graphical representation of space we can see a major and catastrophic flaw in the three-dimensional configuration - see the diagram below (part "C" and "A"). The fatal flaw of the three-dimensional configuration of space is there, where "they" are always putting the reference point (in the corner) which is out of the volume of the considered space or the considered surface. There is no such point in the Universe, where we can put the reference point out of Space! - From sea navigation, we know that to be able to define each point of the map we need to have four directions (see part "B") - East, West, North, and South. They are the four minimum directions (vectors) to be able to define each point of a flat surface when the reference point is inside in the boundary of the flat surface. The same principle is valid for defining any point of space when the reference point is inside the considered space - we need a minimum of six vectors to define it because we cannot measure the dimensions in reverse! – <u>We have to understand that the Reverse Direction is a direction of another dimension</u>! (see the diagram below - part "C").
On the diagrams "A" and "C" are visible, that the reference points are outside – they are on the other boundary of the considered objects - (surface and space volume). In the Universe is no such place, where the reference points

11

can be situated outside! - Everything is inside in the volume of space! So...to define a flat surface the reference point must be into the considered surface! And to describe a surface it is obvious that we need a minimum of four vectors!

But to define a volume of space, when the reference point is inside the volume of space we need a minimum of six vectors - as the well-known directions East, West, North, South, Up, and Down!

I have to repeat it! - **It is the most important fundamental realization of our understanding of the World that the single dimension is a Direction, which cannot be measured in reverse!**

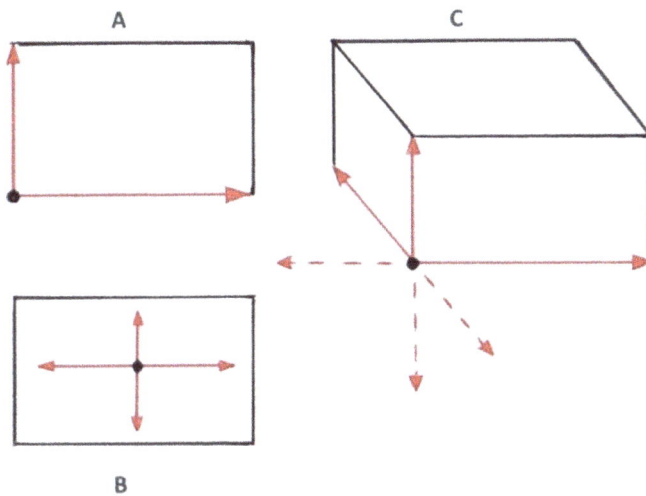

In part "C" is visible that three vectors (dimensions) can define only 1/8 of the total volume of Space and that we need six dimensions to define the total volume of space

Our ancestors were teaching us! - We have drawn millions of maps and our Sea navigation (Surface maps) is based on the principle of Four Directions! - East, West, North, and South! (See diagram "B" above).

Unfortunately, the scientists didn't get notice of this fact and succeed to make a real scientific mess with the assumption for the existence of only Three Spatial Dimensions. - See the diagram "C" where is visible, that three dimensions can define only 1/8 of the total volume of Space because Spatial Dimensions have only one direction and cannot be measured in reverse from the reference point! **The reference point is an incredibly important element in the structure of the Universe.**

12

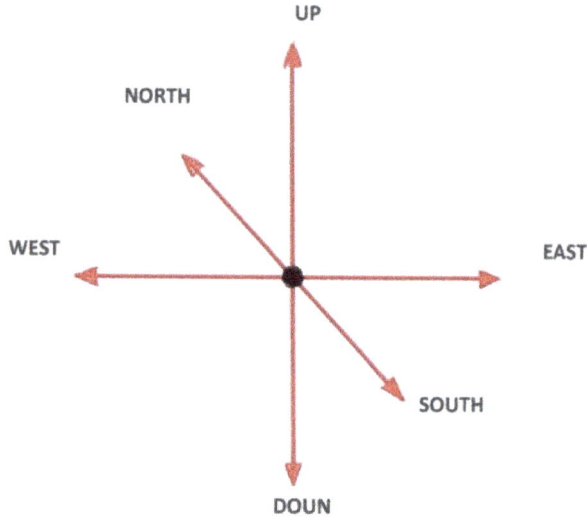

To define each point of space we need a minimum of six vector directions. That's why our ancestors give us the six World directions, but our modern science didn't get notice of it. That's why I am insisting that our ancestors have a better understanding of the World than us.

From the provided examples become evident that our space cannot be three-dimensional, and we need a minimum of six dimensions to be able to have proportional and well-balanced space. Lately, when we start considering the other physical properties of matter we will find numerous facts which are proving that our space is six-dimensional.

## TIME - WHAT IT IS, AND HOW TIME IS PROVIDING THE IRREVERSIBILITY OF THE PHYSICAL PROCESSES IN THE UNIVERSE

Time is a very difficult subject even to be described. Time dilation is giving us the understanding that Time is an energy-related phenomenon and belongs to the physical part of the Universe. So far, nobody has come even close to find and measure the Time energy field, or the physical constituencies and physical property of Time. We are observing that there is a range of limits of knowledge embedded in the property of matter and the laws of physics, which prevents us from obtaining the knowledge and capability, to mess with the order of the Universe. It is more than obvious that these limits are there for a good reason and they are not accidentally inserted! The knowledge of time's property is the next area, where the embedded chain of restrictions of knowledge definitely should be implemented! - And exactly this is the reason - the physical property of Time is to remain a hidden and well-guarded secret of the Universe. The limit of knowledge is preventing us from knowing the exact physical property of Time but is not preventing us from knowing how Time works, and how the mechanism of the irreversibility of the physical processes in the Universe works.

Everybody is talking about the arrow of time, but nobody has yet realized what exactly it means and has been able to resolve the Mystery of Time. Einstein included Time in his theory but wasn't able to explain why and how Time makes the processes irreversible. Scientists are looking into the microscopic property of matter to find there the mechanism of irreversibility. They are calculating and manipulating numbers relating to space, time, dimensions, and are creating bizarre theories, but they are not giving us any reasonable and logical explanations of this puzzle. The invented mathematical formulas with selected and adjusted numbers do not make sense and are a road to nowhere! These formulations are leading only to bizarre theories, wormholes, parallel universes, time travel, and much greater fantasies which are excellent scripts for Hollywood movies, but they are just fantasy without any concrete evidence and do not provide credible solutions and explanations of Time.

I believe that I have found the answer and will be able to unlock the secret of Time mechanism and to give you a good and easy-to-understand credible explanation, based on logical consideration of the available facts, and the well-known relations between the fundamental properties of the Universe. As usual, most of the genius inventions are simple. We are observing the incredibly economical and simple pattern of design in every part of the Universe and its structures. This criterion of simplicity also is applied to the relations of Time, Space, and Matter and also to the mechanism of irreversibility. To understand how the universal physical system works and how Time is producing the irreversibility of the physical processes of the Universe, we have to start with a correctly formulated fundamental physical property and elements of the Universe:

The matter of The Universe is situated not in three, but in six-dimensional space where everything can move freely in all six directions. The volume of our space is proportional to the amount of matter. We have to understand the fundamental difference - that every physical object is three-dimensional, but space is six-dimensional!

We have to realize the difference that Space is situated into the Time dimension, but Time is not dependent on Space and Matter interactions. - This is the unique relation between Time and Space, but they are not physically incorporated! Space and Time are separate physical entities.

The unique property of Time is - that Time is a dynamic directionally oriented progressive dimension. Time directional propagation and the orientation of its energy field are aligned in the same direction.

The difference between Spatial Dimension and Time is that each Spatial Dimension is passive and has only a specific orientation of its energy field. Such property we do not observe in Time Dimension. It looks like that` Time is a uniform energy field with directional propagation only! Time is energy-related and is part of the physical aspect of The Universe! Time is dynamic and can vary, but Space is steady and constant! A well-known fact is that Space is not affecting Time, and Time is not affecting Space eider. These facts are important and are telling us clearly, that Space and Time are not physically incorporated! - And exactly this is the greatly misunderstood concept of modern physics – the relations of Space and Time. - The fact is that Space and Time are part of the same system, but this doesn't mean that they are physically incorporated! - Time is a dynamic directional propagating physical phenomenon that applies to every part of the Universe.

Understanding the nature of Time and the nature of a single dimension is the key to understand the entire Universe!

To be able to explain how time works; I am providing the drawing below, where is also a visual explanation of the difference between three, two, and one dimension. Also, in the drawing, we can follow the actual process and the mechanism of irreversibility, which the single Time-dimension (or the vector of Time) is providing to the entire Universe.

| 3. Dimensions (Our Space) | 2. Dimensions (Flat Surface) | 1. Dimensions (Direction without volume (time)) |

TIME DIMENSION

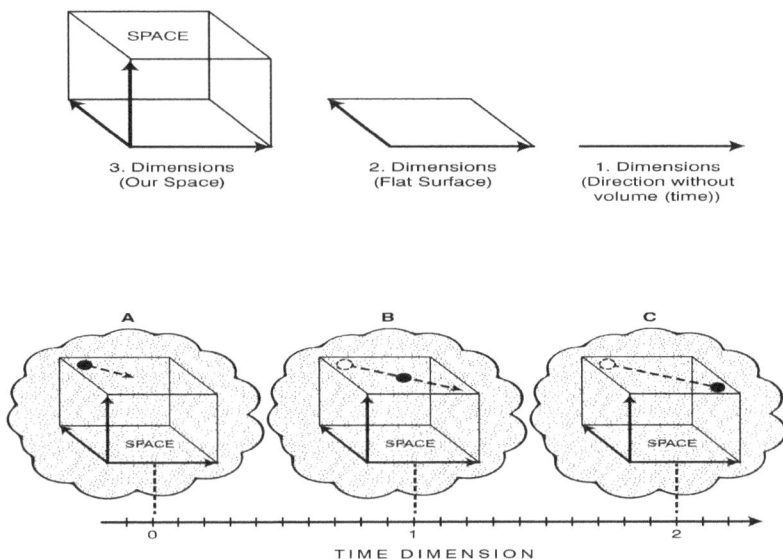

In the diagram above the boxes, "A" "B" and "C" are representing the Space that is moving on the right in the direction of Time. The black object is a particle, which is moving from one corner to another. When the particle reaches the opposite corner (in part "C") the particle cannot go back to its previous (original) position, because its original position is in the opposite corner of section "A" but Space already has moved from "A" to "C" position and the original position of the particle is not available anymore! In the same way, Time is (locking) and providing simultaneously the irreversibility to everything and to all Physical processes of the Universe.

That means that every new second, everything is situated in a different position of the one-directional Time Dimension. The continuously changing (space position on Time dimension) and the irreversibility of Time direction effectively is locking up the physical processes (like ratchet) and makes them fixed and irreversible, because nothing can go back to its original or previous position where they have been in the past! – Effectively, we can say that Time is moving everything continuously in its one-dimensional space direction! The position of objects or the "Reference Point" is the crossing point of Space and

Time. Each Reference Point has its specific Space and Time value!

I just need to mention and make clear that Einstein's concept of physical incorporation of Space and Time is an impossible scenario because Space is passive, but Time is an active dynamic dimension. Space and Time are separate physical entities! - Soonest we realize this, everything starts to make sense!

We have to understand one more puzzle - That the time direction and propagation are not spatial - (because Time is not situated in our space) - Time direction is from the **past** toward the **future** - (terms, which do not exist in Space) Space contains directions, but not Past and no Future! - This is the next important piece of evidence that shows clearly that Time and Space are not mixed together as one substance (space-time) as how the Theory of Relativity insists.  Space is moving in Time, but Time has its separate (independent) way of propagation which is not in the boundary of our space! Future has no spatial orientation!

There is a difference between the nature of the Time dimension and Spatial Dimension - Time has only a directional propagation. Space is steady but has a specific orientation of its energy field, which produces the spin of particles.
 It is hard to comprehend, but this is the best description which we can have, and we know that it is the correct one!

I believe, that was easy to understand what Time is, how Time works, and how Time provides the irreversibility of all physical processes and everything in The Universe.  - How the arrow of Time forms Past, Present, and the Future.  It is a simple explanation of basic fundamental physical interactions which is applying simultaneously to everything which exists in the Universe - from subatomic particles to the biggest structures of the Universe!

## RELATIVITY OF TIME, OBJECTS SPACE POSITION - REFERENCE POINT AND THEIR DEPENDENCY ON TIME

It becomes obvious that Einstein's concept of Time relativity is incorrect, not working and not explaining anything because when we apply his concept we are running into absurd situations as the "Twins paradox" and the "Photon's relativistic time freezing." It is an absurd proposition that in the strictly ordered physical system of the Universe everything exists independently of each other and there is no reliable reference point. – This is chaos!

Ignoring facts and adjusting the absurd results with an aim to "fix" the proposed incorrect concept is not a good way to do science.

I will give you two examples of how the opposite results are used as "undeniable proof" of the incorrect concept of Einstein's Time Relativity. The first example is how Einstein was able to "calculate" and "predict" with absolute precision the Mercury orbit precession. It is well-known that the exact rate of Mercury orbit precession has been known a hundred years before Einstein. (Isaac Newton struggled to explain it). Einstein just made a reverse calculation and has adjusted his theory to the well-known numbers. Even this could be OK if there is not a brutal mathematical deception with the results. The fact is that Einstein's Theory is predicting the exact opposite effect of the observed orbit precession of Mercury. - Theory of Relativity predicts that the Time and the Clock of a fast-moving accelerating object (Mercury near the Sun) <u>must slow down</u>. Further, the Theory of Relativity predicts that the Time of an object near a strong gravitation field (Mercury near the Sun) <u>must slow down</u>.

Instead to sum the two Time slowing down effects as **(-2) + (-2) = -4**
Einstein multiplies them as: **-2 x -2 = +4 ...**And Voila! He cheated out a positive result to "confirm" his Theory with the exact opposite results? - This is no more or less an organized scientific deception and by taking part in it, Einstein is ruining his image as an ethical and intelligent person.

This is the <u>precession</u> (not regression) of Mercury orbit which Theory of Relativity actually predicting

The second example for the opposite results and the opposite predictions of the Theory of Relativity is the "relativistic" GPS satellite clock advancing. According to the Theory, the fast-moving GPS satellite clock <u>must slow down</u>, but in the exact opposite the clock of GPS satellites is going faster than the ground clocks and the satellite clocks must be slow down before launched in space, another way the results will be bizarre.

Anyway, is good to know the state of our current "knowledge," but we have to continue our journey and have to find out what is causing the Time advance of Time dilation.

For easy understanding, we will consider the interactions of Space as a single dimension or single arrow, because in other ways the picture will become too complicated. We can do this because all spatial dimensions are the same, just they have a different orientation, and what is valid for one of them is valid also for the rest of the dimensions.

We have to realize, that every object has a unique position in Space and Time. We need to consider what that means:

That Time is included in defying the position of all objects in the Universe.

I will start the explanation with consideration of two points in space - our position which we can call **"Here and Now"** and another point at a distance which we can call **"There and Now."**

The terms ("**Here**" and "**There**" representing the object's position in Space, but **Now** is its position in Time).

Every object exists only in "**Now**" but each **Now** is situated at a different position of the Time arrow.

Time has only two specific properties: "Now" and the propagation toward the "Future". From our perspective, our time starts from our position and propagates out - toward the other reference points. We know that Time is part of the material structures of the Universe and as a material substance cannot have unlimited speed. That means that when our time starts propagating from our reference point out, the Time needs some time to reach the next reference point. - Effectively, this is the Time difference between two objects. That's why every point of space has a different Time value!

For simplicity, we have to consider Space and Time as dimensions (arrows) that are crossing each other at the "Reference Point".

Correct understanding of "Spatial Reference Point" is very important for our understanding of the fundamental interactions of the Universe.

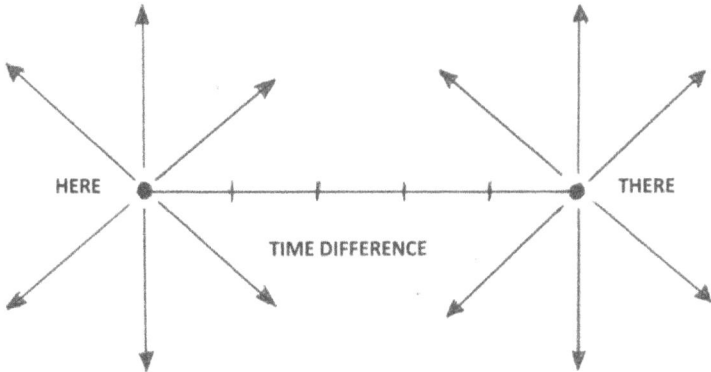

Every physical object has its specific point not only in Space,
but also in the Time Dimension.

**Every point of Space can be a reference point and can be "Now,"** from every point Time and Space is propagating out - toward the future in all directions. That means: - every point in Space has a different Space and different Time values! - (See the graph above)

For easy understanding I will give you an example - if you are standing on the North Pole there all directions are South. If you are standing on the South Pole, all directions will be only North. Same phenomenon we are observing with the two positions on the diagram "Here" and "There". That means that every point of space could be a reference point, where Time is propagating outwards and all directions are toward the future. Every point of space could be "Now" but "Here" and "There" are located in different places of the Time arrow.

From this example, we are reaching the understanding that **every point of space has its unique location because is situated on a specific crossing point on the Arrows of Time and Space Dimensions**. Or in short, the Time is not changing its speed (how the Theory of Relativity proposing) the Time speed is constant! - Just the Time and Space dimensions are crossing each other in different positions of their vectors and different points have a different Time values. For slow-moving material objects, this Time difference is negligible, but for the fast-moving quantum objects as particles or waves which is carrying with them the information for their original local time value, the time difference becomes obvious. For the "Receiver" the incoming waves will carry more advanced time. This is the currently observed in GPS so-called "Relativistic" Time advance of their clocks. This is also what we observing in

Mercury orbit – precession; (not regression) - which the Theory of Relativity predicts. The difference in Time values of different points in Space is because Space is steady, but Time is an active progressive dimension. Time is a progressive directionally oriented energy field. Currently, we do not have any observations that Space, Matter, or any energy field is able to interact and change the Time value. We need to find the specific Time energy density and (Time difference related to distance) to be able to construct a precise picture of the relations and the results of the interactions of Time and Space.

Time and Space are separate physical entities! Space is steady, but Time is active and progressive. - And exactly these separate physical properties are providing the mechanism for the irreversibility and position separation of every object in Space!

In the argument: is the Time absolute, or relativistic, the absolute Time is more credible because if the relativistic Time is ruling our world, the top of the mountains and skyscrapers will slowly drift back and lean toward the west, because of the alleged Time difference provided by the earth's rotation. The most important is that now we understand how Space and Time are constructed and what the relation between them is. Our army of scientists will have fun finding the rest of the details.

We are measuring Time with our clocks, but we do not measure the length of Time! We are measuring its' speed only! The speed of Time is related to the intensity and momentum of its energy field!

I am afraid that I will disappoint many people, but we are coming to the point of understanding that the fantasies for Worm Holes and traveling to the past are baseless and impossible! It is prohibited by the laws of physics because Time is irreversible! **Time is one-directional by its physical nature of being a One-dimensional vector without beginning or end!**

## HOW TIME APPLIES TO POSITION, VELOCITY, AND UNCERTAINTY AS A DEFYING FACTOR

From the above chapter, we learned that the Time mechanism is applied simultaneously to all parts of the Universe and is providing the irreversibility of all physical processes. (See the graph page 16)

To be able to understand the principle of uncertainty, we have to understand the unique futures of Space and Time in defying the object position and velocity. We know and are easy to understand that objects can have a different position in Space, but currently is unknown, that **the different positions in space also are different positions in the Time dimension**. - (Like two vectors that are crossing each other in different places).

We are defining the velocity of an object as the Time difference between its consecutive positions in space.

The object's position we are defining as its simultaneous present position in Space and Time. - It is obvious, that to measure the velocity of an object we need to know two values of Time in two consecutive positions and their difference, but to define the exact position of the object, we have to know the precise value of time in a precise point of space! (See graph pg 20) These two demands are conflicting because we can know only "One" Time value at the time, and we cannot apply simultaneously three different time values on the same object, (at the same time) because if we use precise Time, we can define the precise position of the object, but we are losing sight of its velocity. And the opposite - if we apply to the object the time difference between the two points in order to define its velocity, we cannot apply the precise time to determine simultaneously its position! - We reached an understanding that we cannot apply three different Time values on the same object **at the same time!** - This is the Time role in the "Puzzle" of the uncertain velocity and position. Modern science doesn't know that Time is involved in defying object's position. The fact is that we are dealing with Time to define the velocity and the position of the objects is stripping us of any possibility to know simultaneously the velocity and the position of the elementary particles. This is just a matter of technicality and also is one of the cleverly inserted "Limits of knowledge", because with aid of a supercomputer we will be able to design and predict the future of some part of the Universe if we know the position and velocity of the particles there.

### UNDERSTANDING SPACE

The final configuration of space properties currently is beyond our reach. Fortunately, the interactions of the fundamental forces and elements of the Universe are giving us a good base for a basic understanding of the fundamental structure of our Space - that Space is a medium capable to provide conductivity to all forces and radiations in the Universe. Space is containing all known to us forms of energy and energy fields. To understand

Space, we also have to consider the other two physical components of The Universe (Time and Matter) and their fundamental relations to Space.

We have to start with the understanding that space is an energy-related physical phenomenon. We have to know, that for a specific amount of matter (energy=matter), there is allocated a specific (proportional amount), or volume of Space - evidence for this is the homogeneous structures of the Universe on a grand scale. Further, in the next chapter, we will see the reason why Space and matter are in strict proportional dependency! - This relation of Space to Matter means only one thing - that Space is a constant and unchangeable fundamental component of the Universe because we know that the amount of matter is constant and is not vary! The constant volume of Space is providing the stability of the physical system of the Universe. This structure is providing the condition of a <u>closed physical system of The Universe</u> – (regardless of is the Universe is finite or is endless) - the energy and all physical processes have nowhere else to go! - This is the key factor behind the finely balanced matter-energy exchange mechanism of The Universe, which we will consider in a later chapter. The closed physical system of The Universe is providing conditions for the endless cycle of energy-matter exchange and all the universal structures to be continuously recycled. <u>This provides the eternal dynamic existence of our Universe</u>.

We have to realize, that Space is a physical medium with specific physical properties, and the ability to propagate waves. Space is a perfect physical medium, with the conductivity range from zero to superconductivity, dependent on temperature, matter density, and the nature and intensity of the fields involved. Space is acting as a real physical medium containing its own enormous energy, which provides the different energy levels for the attractive mechanism in the Universe. Space is giving finite speed to traveling particles and allows propagation of the electromagnetic waves - (Light). The observational evidence for this is all around us. The revealing fact that Space is a physical medium is the fact that the passing waves and particles are losing energy – (That's why the Universe is not blindly luminous). We know that at every point of the sky is a star, and if the light is not depleted, the sky will be blindly luminous, like we are surrounded by a sphere of billions of blinding suns! This is the reason for the apparent Red-shift of the distant stars. Science has to find the exact rate at which gravity and electromagnetic fields of the Universe are depleting the light energy and what role Space is taking in this process. It looks to me that Space has a slight resistance to the passing waves and this resistance is taking part of the wave energy and is transforming it into thermal energy. Good observational evidence for the light depletion is the Cosmic Microwave Background Radiation. This radiation is the shine of the distant galaxies of The Universe and is coming from a distance beyond our observational range of (the visible Universe). The energy of this light has been depleted below the visible range of the light spectrum - to the range of microwaves! (See the diagram on page 40)

The assumption that our Space is three-dimensional is not credible! In the previous chapter, we have considered the physical nature of a single dimension, which is a progressive, dynamic direction, - without a beginning or end and without volume! The assumption of the leading scientists that the description of the World has to be in the form of mathematical formulas is not correct eider. For example, mathematicians are describing surfaces as two-dimensional phenomena, where two vectors can define any point of the surface. The difference between this mathematical assumption and the real world is that the mathematicians putting the reference point of their vectors always in the corner (which is out of the measured surface!). But in the real world (as a map) we and our reference point always are on the surface of the map. - To define a surface, (or map) we need not two, but four-vectors or directions – <u>east, west, north, and south!</u> – This is the difference between baseless mathematical assumptions and the reality of the physical world. The same fundamental mistake is made with the assumption that we can define the volume of Space just with three vectors only because the mathematicians again putting the reference point of their vectors in the corner of the measured box, - which is out of the defined space volume!

I am sorry, but there are no corners in the Universe! - The Universe is all around us! Every point of the Universe is a reference point for defining the space around it! - There are no corners in space <u>and there are not three, but there are six spatial directions!</u>

The fatal mistake of the current three-dimensional configuration of space is the wrong assumption, that from the reference point "they" can go backward in the opposite direction of the three vectors to define the rest of the space. This is impossible because each Space Dimension has only one direction! The act of measuring in the opposite direction means that <u>you are measuring a different dimension with an opposite direction!</u>

We have to realize, that each direction starts from the reference point and is separate dimension! Each reference point in space is "Here and Now".

We, humans are perfectly designed to feel and sense the real world. There is no better and more credible description of the World than the picture, which our senses providing to us. We know that the Cosmos is uniform in all directions; We know that there is only one Time Dimension; We also know that the World has six directions: - for example - 'West' is just a direction, it has no volume, no beginning, no end, but is a real direction, it is existing, and we know that we can travel in this direction! We also know, that there are another five space directions! East, North, South, Up, and Down. We wouldn't know for the existence of the six world directions if they really do not exist! We have to realize also the fact that each world's direction exists because it is representing a separate dimension. - They are six directions, we know them, and prove for it is that we can travel in each direction from each point of space. I will repeat again, - To define the space around us, we need not three, but six directional vectors because three space directions (Vectors) are not enough to define the whole volume of space. In the diagram below, it is visible that three dimensions are able to define only 1/8 of the total space volume! Three-dimensional scenario will produce a disproportional configuration (distortion) of space and an uneven continuous expansion because three vectors are defining only 1/8th of the total volume of space, and their dynamic momentum remains active and cannot be restricted of continuous progress. - (See the diagram on the next page)

Now we have to consider the subject of space dimensions and how they are incorporated together:

The correct understanding of dimension is that the dimensions are progressive directionally oriented energy fields, which have dynamic directional propagation momentum and specific orientation of their energy field (as directionally oriented vibrations). The correct understanding of the physical property of Space requires a total reconsideration of our understanding of Space, Time, and the Universe. The current official assumption for our space suffers a lack of understanding of what really the dimensions are! – That the dimensions are directions, which cannot be measured in reverse! This realization is explaining the origin of the fatal flaw of the three-dimensional assumptions for our space. Our space cannot be three-dimensional only, because three dimensions are not representing the total volume of space, and will produce a fatal continuous space distortion! We have to understand the fundamental difference, that the physical objects are three dimensional and they are defined by three vectors <u>with defined length</u>, but, <u>space cannot be formed and described by defined length vectors, because space dimensions have one-directional</u> **progressive energy-momentum** <u>and because there are six separate space directions</u> – East, West, North, South, Up, and Down, <u>and for each space direction to exist it must be represented by a separate dimension!</u> - It is vital to understand the difference: that three-dimensional vector configuration is defining only 1/8th of the total volume of space! And in the three-dimensional space scenario, <u>the progressive dynamic momentum of each dimension</u> **will remain active**! This unrestricted dynamic momentum will distort and (expand) the 1/8th space defined by three active separate dimensions, which will continue expanding forever, reducing space energy density and diminishing the matter at a very fast catastrophic rate! **Three active dimensions <u>are not enough</u> – they cannot produce a steady, uniform, and balanced space, which we are observing and enjoying!** <u>In order to have a steady and uniform space,</u> **the active directional propagation momentum of each dimension must be canceled!**

And the active momentum of each dimension in our Universe is canceled with the active momentum of an oppositely oriented dimension. - (<u>East</u> is canceling with <u>West</u>; North with <u>South</u>; and <u>Up</u> with <u>Down</u>). - <u>The dynamic momentums of our Space dimensions are canceled, but the specific orientation of their energy fields remain</u> in a steady-state and preserving their spatial orientations! - (See the drawing below and page 29)

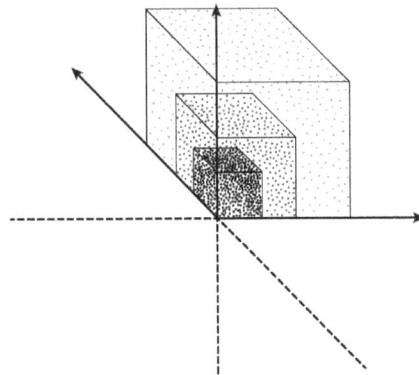

3D vectors producing only 1/8 of Space volume
and continuous (catastrophic)distortional expansion

This is the configuration, which is producing the steady; define volume of space full of enormous energy. - It is the uniform space, which we are observing and where we are living in. - It is a World of Energy!

I believe that the explanation of Space was easy to understand and the logic and simplicity of space construction are giving us the confidence to continue our journey forward and to apply the principles of this structure to the rest of the elements of the Universe. You will be surprised to see how easy and simple is to understand how the Universe is constructed and how works when we are using correctly defined fundamental elements.

Some could ask do I have any evidence in support of my theory? The answer is yes! - There is not just "any" evidence, but there is millions of evidence all around us, we just have to know where to look for them.

I will give you some examples and I will start with Time - we know that Time is a single dimension and we know that we have only one Time. Nobody asking proof for the existence of Time, and nobody suggests that there is more than one Time! That's why we have only one watch on our wrist. If we have more Time dimensions, we will know this and will need to have more watches on our wrists. - It is the same situation with space dimensions (or directions). Same as our knowledge of Time, we know that there are six space directions, and I don't have to prove this fact! The fact that we can travel in six different directions from each reference point of our space, is undeniable proof for the existence of six single dimensions of our Space. If there are three dimensions only, we will have no clue about the existence of six space directions - we will know only three directions! We are encounter only what is included in our physical world! - No more, No less! - It is simple to understand, that - we won't be able to travel in six directions if our space is three-dimensional only! Full stop! - This is my answer to the question - do I have any proof for the existence of six space dimensions. The proof for it is printed on every map, every compass, and GPS, where it is visible the four separate surface directions – East, West, North, and South. The existence of the other two directions - Up and Down, everybody knows them, and I don't have to prove their existence. - (To define any point on a surface, you need four vectors, but to define and to form space, you need a minimum of six vectors, which represent the six spatial directions!). Space has no special reference point in it. – Every point of space is a reference point – (HERE and NOW), where from our perspective the Time and Space dimensions are crossing and the six dimensions are beginning and propagating out in all directions. Our unique personal point of existence (no matter where we are situated) always providing us with the ability to travel in each six space directions!

The next crucial point of understanding Space is its relation to Time! The assumption of the current 'Standard Model' that Space and Time are physically incorporated in one substance called - 'Spacetime' and forms four-dimensional space is absolutely incorrect! Einstein's 'spacetime' is physical incorporation of four single dimensions, which is a real physical mess! It is not working and cannot produce, or explain anything!

An additional active dynamic dimension as (Time) incorporated in the well proportional and steady space will create a fatal distortion of any shape and any object! Space and Time are separate physical entities and the relation between them is different than the currently assumed 'Standard Model'! Space and Time are not physically incorporated! They just share the same volume! **Time is dynamic, but Space is steady.  Steady and Dynamic dimensions cannot be physically incorporated!**

I believe that the explanation of space was easy to understand and the logic and simplicity of space construction are giving us the confidence to continue our journey forward and to apply the principles of this structure to the rest of the elements of the Universe. Don't be surprised, will be amazingly easy and simple to understand how the Universe is constructed and how works when we using correctly defined fundamental elements. Later, when we are considering the atomic structures of matter we will find that the atoms are constructed also on the base of a six-dimensional configuration and that electromagnetism also is based on two sets of dimensions which different spatial orientations providing the electrical polarization of plus and minus.

## SPACE DIMENSIONAL MOMENTUM CANCELLATION AND ORIGIN OF MATTER IN THE UNIVERSE

Regardless of our different views and beliefs, we still have to recognize that the Universe has been created at some point in time in the distant past. When we put aside the impossible scenario of Big Bang theory, and the miraculous religious explanations, in front of us remains only two options - Spontaneous formation of the Universe, or to be created by super-intelligence, (which we normally associating with God).  We will not consider which scenario is the correct one. We just have to realize, that in both cases somehow, the building blocks of the Universe have to be put together in good order to works. - Exactly this will be the subject of our consideration - what will happen when the six active dimensions have been put together to form our Space?

We reach the understanding that every object of the Universe is suspended in space and can travel in six different directions! – East, West, North, South, Up, and Down! All those six different directions are representing the six single space dimensions of the Space of our Universe!

We already reach an understanding of the nature of a single dimension - That single dimension is a one-directional propagation of directionally oriented energy field. We have formulated a single dimension as "**Active**" because it contains two active properties - **directional propagation** and **orientation of its energy field.**

We had discussed the difference between Time and Space, - that Time dimension is a uniform active dimension, which has active directional propagation. Space is an assembly of six passive dimensions. Space Dimensions do not have active dynamic propagation. Space Dimensions have only a specific orientation (as vibration) of their energy field. So... where is gone the original dynamic momentum of the Space dimensions?

The answer to this question has enormous significance for our understanding of the Universe. It is easy to comprehend, that when the six spatial dimensions have been put together, three of them will have an opposite dynamic direction to the other three Dimensions. (See the graph below)

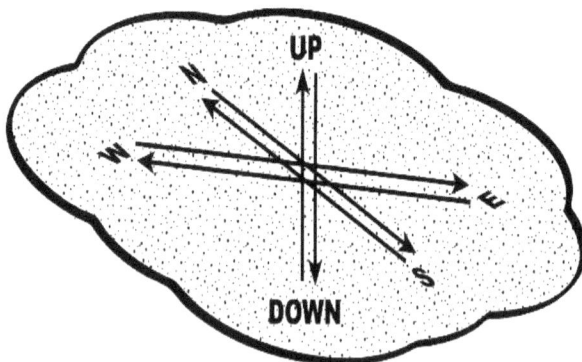

The 3 pairs of opposing space dimensions are canceling
their dynamic momentum and producing
Uniform steady space and releasing energy = (matter)

From physics we know, that when the same size objects or energy fields with opposing directions meet, they will cancel their dynamic momentum, and the kinetic energy of this confrontation will turn into heat.

In our case, the dynamic momentum of the Space Dimensions is absolutely the same! - That means that when the two opposite oriented active dimensions meet, they will cancel their dynamic momentum and will become steady dimensions. The result of the cancellation of the dynamic momentum of the Space Dimensions will produce heat. We know that any form of energy, including heat, is a mass! This realization giving us the understanding that the result of the cancellation of the dynamic momentum of the six active Spatial dimensions will be - Six Passive Dimensions preserving the specific orientation of their energy field and **heat** which is equal to **Mass**!

This is one of the great applications of the correct understanding of the nature of a single Dimension. It is a great example that when we using the correct property of the fundamental elements and apply them to the model, everything is falling in place with ease, everything fits and makes sense!

Let summarize our finding: In the six-dimensional configuration of Space, we are observing three pairs of single dimensions with opposite direction! - This is very important because this directional confrontation providing the required conditions for the cancelation of the dynamic momentum of the opposing each other spatial dimensions.  As a result of the cancellation of the dynamic momentum of space dimensions, they are forming a normal steady volume of space, where each spatial dimension is losing its dynamic momentum but preserves the spatial orientation of its energy field.

(For example - It is similar to the scenario where two identical spinning marbles collide and are canceling their kinetic energy but still preserving their rotational spin).

Understanding the space dimensional momentum cancellation is fundamental for our understanding of the World because exactly this phenomenon is explaining the origin of all the matter of our Universe!

This is the mechanism, which provides a strict proportional ratio between the volume of Space and quantity of Mass in the Universe!

The strict ratio between Space and Mass is giving us an understanding of another important subject - (the speculation of Space expansion). We know with absolute certainty that the amount of matter of the Universe is constant!

The law of Physics is stating that you cannot add or take away even one atom from the physical system of the Universe. This knowledge is giving us the understanding that <u>if the matter of the universe is constant and is not varies; the volume of Space also must be constant and is not varying</u>! - This is simple, but is a fundamental understanding of the nature of the Universe!

This explanation is based on real physical interactions and is much more credible than the explanation that "<u>the Universe comes from **Nothing** for **No Physical reason</u>**".

Some could ask where the antimatter in my model is? My answer is that this is a purely speculative question unrelated to my model. First, the appearance of antimatter in the particle accelerators is irrelevant to the conditions of energy cancellation of spatial dimensions. And second, this question could be raised only, if it is performed an experiment with the cancellation of the energy of two opposite dimensions and the antimatter is present in the result of this experiment. - Most likely will be that the cancellation of space dimensional dynamic momentum will produce heat or neutrinos. Heat will never turn into antimatter. So far, the neutrinos are known as the smallest and the most abundant particles in the Universe. The neutrinos and antineutrinos are not annihilating each other and are able to change the "flavor" (or change into each other). - This could explain the speculatively assumed "lack of antimatter" in the Universe!

My explanation of the origin of matter in the Universe is based on the real physical properties of the fundamental elements of the Universe. It is logical and is not violating the Law of Physics and also is not an assault on our intelligence, how is the case of the "Modern Science" which claims that "The Universe comes from Nothing." Despite that, the Law of Physics states that you cannot produce energy.

## WHAT IS ENERGY AND ENERGY FIELD?

We are using and defined many forms of energy in our everyday life – as Kinetic energy, Potential energy, Thermal energy, Electromagnetic energy, Strong and Weak  Nuclear energy, Pressure, Magnetic, Pneumatic, and named... We know that the fundamental form of Energy must be in one form only and to apply to all known different manifestations of Energy. Unfortunately, we are living in a total lack of understanding of the

fundamental elements of the World. Current science cannot defy even the basic components of the Universe as - Space, Time, Matter, Information, Consciousness, Law of Physics, Electromagnetism, neither their origin.
The same situation we have with defying and understanding of Energy and Energy fields. Modern science has no answer to these questions.
I will try to change this situation and will explain what actually the basic are – or the fundamental form of all known to us forms of Energy is. I will start directly with the definition of what Energy is and then will give a detailed explanation: - Energy is a disturbance of the equilibrium of Space.
A few pages above we have discussed the structure of Space. For simplicity and to be easy to understand the concept of Energy, I am providing a graph below with the micro-layers of Space dimensions. (or the fabric of Space)
We can consider Space as a balanced energy field. Any disturbance, force, or movement there, will disturb the equilibrium of the "space pressure" and will create increased (or decreased) pressure in the point of interference. This will push (distort) the boundary borders between the layers of Space. The Law of Physics dictates that the pressure equilibrium must be restored!
(see the graph below and the graph on page 40)

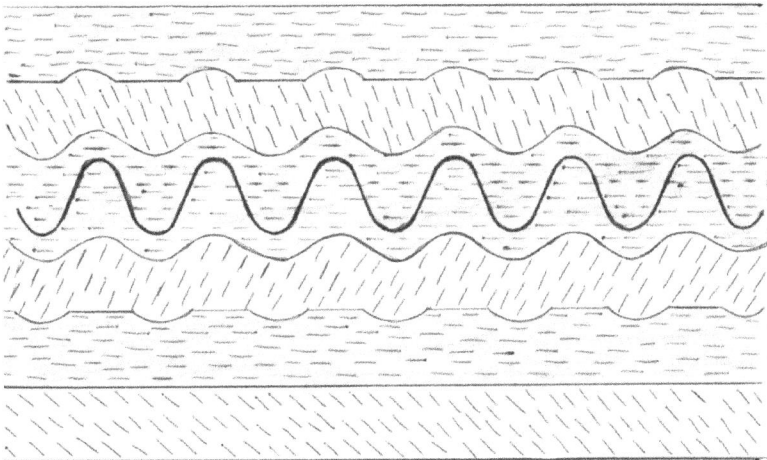

The tendency of the Physical System to restore its Equilibrium is the fundamental "tool" for the manifestation of Energy! The Space equilibrium is the observed (neutral state of Space) called "Zero-point energy" The potential disturbance of the Space equilibrium and the tendency of Space to restore back the equilibrium is the manifestation of Energy. Theoretically, Space could have nearly unlimited energy because when you distorting more and more space layers, the stronger response the system will apply to restore its equilibrium.

There is one simple example. - If the electromagnetic wave is traveling in one of the layers of space, the pulsating sinusoid of the electromagnetic wave will expand or contract the boundary of the conducting layer and the boundaries of the neighboring layers. The interference (the distortion) of the equilibrium of Space is the fundamental mechanism, or it is the basic form of all kinds of Energy.

Now, when we manage to define what Energy is, will be very easy to define what an Energy field is. - Energy Field is an area of Space with disturbed equilibrium. We know that this definition is correct because all forms of energy are a manifestation of the disturbance of the space layers' equilibrium.

## THE ATTRACTIVE MECHANISM IN PHYSICS AND THE UNIVERSE

The inherited incorrect-expanding and open space model of The Big Bang theory and Einstein's weird assumption for space curvature create a real physical mess and are stripping us of the chance to understand the real nature of Gravity and what the attractive mechanism and forces are. Those fundamentally incorrect configurations are leading to the creation of endless weird mathematical models and theories, which do not represent reality, do not make sense, and are not explaining anything. The assumed physical properties as - bent space, gravitons, gluons, and the "discovery" of "Gods favored -" Higgs boson is a dead-end road!

In reality, there is nothing mysterious or complicated to understand how the attractive forces and gravity works. It is not difficult to comprehend how the Universe works if we keep our feet on solid ground and using the law of Physics, proven facts, observations, and logic to build a correct model and correct configuration of the fundamental building blocks of our Universe:

We have to realize, that in our physical world of particles and waves, there does not exist any substance, force, or mechanism, which can provide physical attraction! - We can push and propel objects with particles, but we cannot pull them back or attract them with particles! – It is impossible! We have to realize and accept this as a fact!

The claim of the academic establishment that this is our "best explanation" is not a reason to accept such nonsense. We need to use our knowledge and logic to find out how physical attraction really works.

So... how with propulsion only in our arsenal, we can have an attractive force? It won't be very difficult to figure out if we continue to use correctly defined fundamental elements, their interactions, and healthy logic:

The science is recognizing that vacuum (or space) is filled with energy - a load of energy! (The explanation is in 'Understanding Space') There is a continuous argument for the strength of this energy, where the predictions vary between $10^{-9}$ to $10^{113}$ joules per cubic meter. The last figure is enormous! Some scientists are stating that in one cubic centimeter of space, there is stored more energy than the energy of all matter content of the Universe! Even if just a fraction of this claim is correct, the closed physical configuration of The Universe will provide conditions for this energy field of space to acts as a uniform energy pressure surrounding all material objects and act as a pivoting point for the attractive mechanism and forces. In a pressurized environment, the reduction of pressure between two objects is forcing them to move toward each other. A similar situation we observing in space - energy cancellation and imbalance between any two material objects will deplete slightly the density of the space energy field between those two bodies and the energy imbalance will start pushing those two bodies toward each other, similar to the two objects with opposite magnetic polarity shown on the picture below.

The Law of Physics postulate - that energy and matter are moving from a state of higher energy level towards a lower energy level. – This is the key behind the existence of the attractive mechanism in the Universe. It is the same mechanism of attraction and repulsion of all forces of the Universe!

The forces of Gravity and Electromagnetism are acting in the same way! In the picture below it is visible that the energy field surrounding the two objects is uniform in all directions and acts as energetic pressure on the objects. The opposing polarity of the objects is canceling the energy field between them, and this creates an energetic imbalance- (as energy vacuum). As I stated above, the law of physics dictates, that the outer energy (pressure) will push the two objects together! – This is what we are observing with magnetism. This is the Universal Principle of Attraction and Repulsion! There is no "Mystery" involved. This is a simple physical process and is the common mechanism for all forces in the entire Universe!

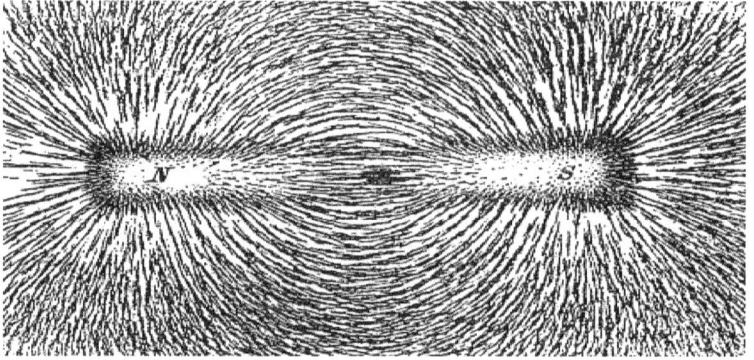
(The opposite charges are canceling the energy field between them and producing "energy vacuum")

This is a well-known physical process where the natural tendency is to move from a higher energy field towards a lower energy level. This is the mechanism, which explains the "mystery" of gravity, and all attractive forces in the Universe! The electromagnetism and nuclear forces are working in the same way – they are a disturbance (or local imbalance) and cancellation of the energy field between two bodies!

One of the best pieces of evidence in support of the concept of energy deficit as the mechanism of physical attraction in the Universe is the well-known mechanism of chemical reactions. We know that the chemical bond of two atoms is a <u>missing electron</u> and they are sharing a common electron of the other shell. From electromagnetism we know, that expelling an electron and sharing a common electron, will make the two atoms positively charged and this will repel them away from each other. Contrary to our intuition, the missing electron is acting as an <u>energy deficit between the two atoms</u>, which is keeping them together!

Maybe I will disappoint my readers with my short and simple explanations, but it is not my fault! I have to say again when our concept is correct everything is simple and the elements are falling in their places very easily without the necessity of complicated mathematical formulas to justify the obvious!

With correct configuration becomes easy to understand how the force imbalances are producing attraction! - This is the answer to the elusive question of what the nature of the attractive forces is, and how Gravity really works!

# SOLID-STATE OF PARTICLES AND WHAT IS DETERMINING THEIR SIZES

We know that the particles exist in two states - as a wave, and as solid particles. Modern science has some understanding of what the waves are but has no good understanding of what actually is the "solid-state" of particles. From the graphs above in the chapter of Light, you can have a better understanding of the nature of the waves, which is harmonic propagation of a line of consecutive points of energy excess and energy deficit in space. We have to understand that there is nothing "Solid" in our Universe! Everything material is made of a concentrated form of cleverly balanced electromagnetic forces. The "Solid-state" of particles is not making an exception. - They are a tight and well-balanced concentric configuration of standing waves around the center of space energy deficit point. The mechanism of formation of the "Solid-state" of particles is not well understood by modern science. The collapsing of the "Wave state" of particles into a "Solid-state" is very complicated because there is involved also the phenomena of Consciousness, the Law of Physics, Time, and Quantum Information. In addition to the mentioned phenomena in the formation of the "Solid-state" of particles significant role is playing the energy level, polarity, and space orientations of the waves involved in the process. It is a very interesting question - what is determining the size of the elementary particles, and why the atomic particles are extremely small? What is stopping them to grow bigger as planets and stars? Modern Physics has no answer to this question. We have to see if our model has an answer and can it explain this great puzzle?
We have an understanding that the basic formula or the configuration of a subatomic particle is a standing wave oscillating around the energy deficit center. From this configuration will raise a logical question - why the elementary particles are small and what is restricting them not to increase their size and form a black hole and structures of unlimited proportion? - Modern science has no answer, but we do because we are possessing correctly defined fundamental structures, where everything fits perfectly and the explanation always is simple, elegant and we don't need complicated mathematical formulas to understand it. - That's why the ancient philosophers have a better understanding of the structure of the World than us because instead of using mathematical formulas, which have no meaning, they were using observations and Logic!

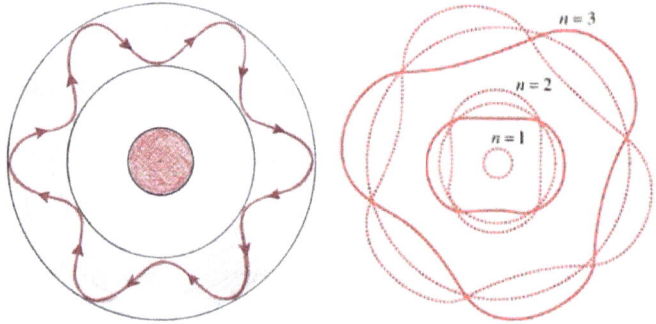

Graphical representation of elementary particle

We know that the fundamental property of the elementary particles is their spatial orientation or (spin). The spin of a particle is its "identity." To find the answer to our question we need to consider carefully what exactly this means - The spin or the spatial orientation of the particle is representing the spatial orientation of its standing wave. But the spatial orientation of a particle is dependent on the orientation of the dimension where the particle is produced! The thickness of the micro-dimensions is determining the size of the elementary particles. – For some particles to have one define orientation, the particle must be small enough to fit in the size of one particular dimension. This is the key behind the size of the elementary particles. The standing wave of a particle can have a combined orientation of two or mostly of three non-opposing spatial dimensions. The particle cannot occupy four, five, or six spatial layers, because the opposite orientation of the dimensions is canceling the standing waves with opposite orientation. (See the graphs below)

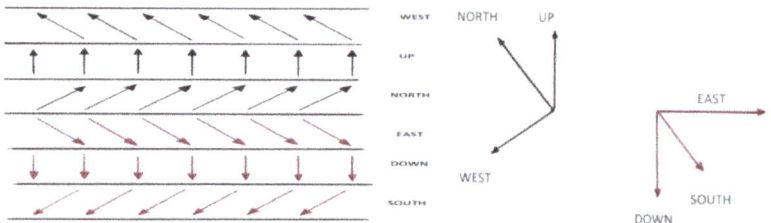

The fine space layers and their opposing orientations

From the graphs above became easy to understand that the back and red dimensions have opposite spatial orientation and no particle can co-exist simultaneously in two opposing sets of micro-dimensions. The thickness and different orientation of the layers of space dimensions are limiting the size of the elementary particles and are not allow them to grow bigger and to exceed the boundary of their spatial confinement.

The size of the particles is a good indication of the thickness of the layers of the micro-dimensions of Space. Knowing the thickness of Space layers will give us a very good explanation for the weird behavior of the micro-world of Quantum Mechanics, which we have to correct and re-defined.

## LIGHT

Light is the unifying link between the thermal, kinetic energy, and electromagnetic property of The Universe.

Light is coming in two forms - as particles (photons) and as Electromagnetic waves. The confusion in the accepted model for the property of light in the "Standard Model" of physics is coming from the fact that Photons are not composite particles and are regarded as fundamental particles. The officially accepted view is that Photons are mass-less. We have to come down to reality and to realize the fact that "Something" cannot be made from "Nothing" and if Photons exist in form of particles capable to apply pressure on the illuminated surface, this "Pressure" is an indication of the mass of the coming Photons. - Photons have momentum, and momentum is the combination of mass and kinetic energy! We also observed, "Gravitational Light Redshift," which means only one thing! - Gravity attraction is depleting the energy of the photons. Gravitational Lansing also proving that gravitational attraction is altering the path of the light. If Photons are mass-less the gravity will not have any effect on them!

We are considering Photons as fundamental particles because Photons are not composite particles and the photons can disappear without a trace - something, which the composite particles cannot do.

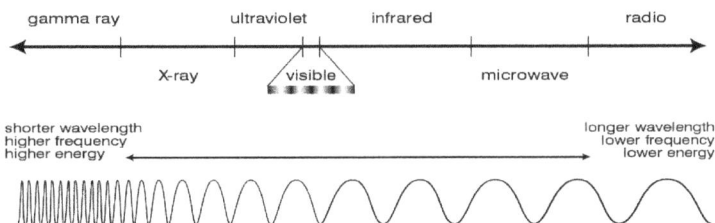

So what actually is the nature of Light? I will start the explanation with the wave function of the light: The simple definition which I can provide is that the Light is a traveling wave through the medium of Space. (See the diagram below)

This definition is not too far from the officially accepted, but in reality, it is fundamentally different. The difference is coming from the involvement of Space Medium. In mainstream science mentioning "Space Medium" is Taboo! - Again we have to come down to reality and realize, that there are no such things as "Wave Traveling in Nothing". The basic nature of wave is a "Harmonic Disturbance of a Medium" and we have to accept it and stick to the reality regardless of how uncomfortable it is to some "established" theories.

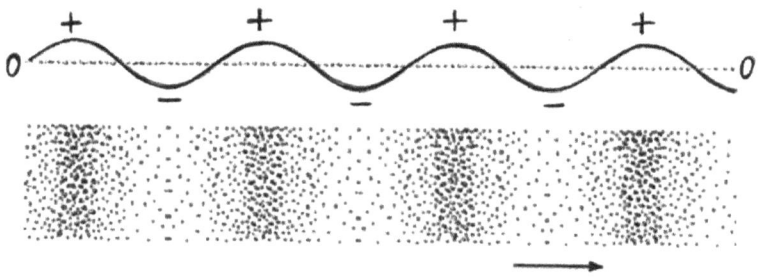

Space is a pressurized medium, but for the traveling wave is acting as a neutral medium (the dotted line "0")or (zero energy level). The ridges (+) and the valley (-) of the wave are energy excess and energy deficit acting as pressurized areas and vacuum points. Both - positive and negative points carry an equal quantity of positive energy.

The traveling light wave actually is curving the boundaries of the space dimensions and creating local energy excess or energy deficit in the points of interactions with the boundaries of the layers of Space Dimensions which we can call a Medium.
I am providing a graph wherein a simple way has represented the interactions between Light waves and the layers of the Space medium. This is the fundamental diagram, explaining the interactions between Space and Waves. - (See the graph below)

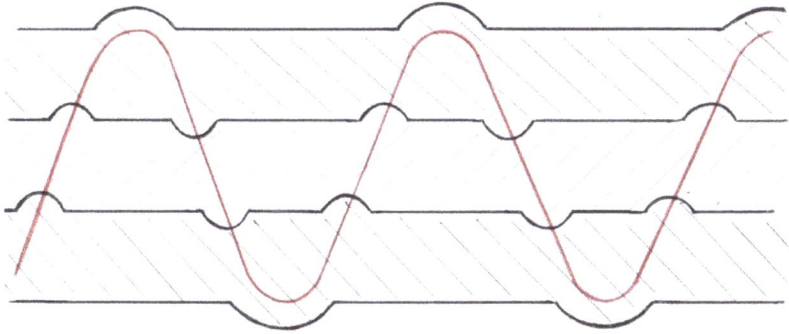

This is how the traveling waves create a distortion of the layers and boundaries of space dimensions

The energy of the light wave is corresponding to its frequency. It is easy to understand that the higher frequency will have a bigger amplitude, and in a certain length of space, will deposit more and stronger disturbances of the equilibrium of Space Medium.

Now, on the solid-state of Light - Photons: In the previous chapter, we have defined that the solid-state of an elementary particle is a point of space energy deficit surrounded by a standing wave. This also is the configuration of the "Solid State of Photons". Photon is just an ordinary fundamental particle. On the question of how the electromagnetic field has direction and spin (the right-hand law) the answer is that the photons have direction and spin too, and are interacting with multi-directional medium (Space). - In such dynamic interactions, the normal outcome is that the result also will have direction and spin. How works the mechanism of transformation of the traveling wave into the concentrated form of a particle? Modern science still has no answer. It is a very complicated process to be understood because there are involved many different factors, which our science is not willing to address yet. - The Law of Physics, Consciousness, Information, Space dimensions, and orientation. All these elements are involved together in quantizing and defying particles and waves. I wish to be able to explain the mechanism of wave functions collapse a bit better, but at least I dare to reveal to you which phenomena are involved in the formation and functioning of the Physical order of the Universe. I suspect that in the collapse of the wave functions the phenomenon of the "limit of Knowledge" also is involved to prevent us to play with the energy-matter content of the Universe.

# ENTANGLEMENT AND THE NATURE OF THE ENTANGLED PARTICLES

The phenomenon of particle entanglement is a puzzle for modern physics because the behavior of the entangled particles disobeys every rule of physics. There is no explanation of how and why the physical properties of some particles are in strict dependency on each other, what the physical link between them is and what the nature of the instant informational link between them is. It is well-known, that if you change the state of one of the entangled particles, this will affect in instant the physical property of the other particle no matter the distance between these particles, even if they are light-years distance from each other. It looks like that the particle entanglement is disobeying the Law of Physics...or we don't understand something on a fundamental level?

I believe that will be easy for me to explain what is going on because we have a correct model of the fundamental elements of the Universe.

The "Standard Model" of physics is based on mysteries and puzzles, but all these mysteries are just a result of wrongly selected fundamental principles. The "Puzzle" of entanglement is coming from the adopted materialistic philosophy. Modern science is refusing to recognize the existence of non-material phenomenon as Information, the Law of Physics, and Consciousness. In the case of information, they are giving information materialistic property. A good example is the so-called "Information Paradox" wherein the imagination of the academic science, the gravitational attraction of a Black Hole should hold the information of the fallen inn particles inside and will not allow the information to get out of the Black Hole. Even a schoolboy will be able to tell them that gravity cannot affect information! - Information is not material substance! Information is a song, melody, story, mathematical formula, plan how to build something! Information is a real phenomenon, but is not part of the material aspect of the Universe! The speed of the material parts is limited, but the speed of the non-material components is not restricted. - This is exactly where academia is getting wrong!

Back to our subject - It Will be interesting to find how our concept of the structure and property of the Universe applies to this phenomenon and how our concept will be able to answer this puzzle? Probably many readers will be disappointed of how easy will be the explanation of this incredible puzzle when we have a correct understanding of the fundamental elements of the Universe:

I will start the explanation with the example of a (normal) single particle. Single-particle is an inserted quanta of energy into one layer of micro-dimension. It can exist as a solid particle, or as a wave. (See the graph below)

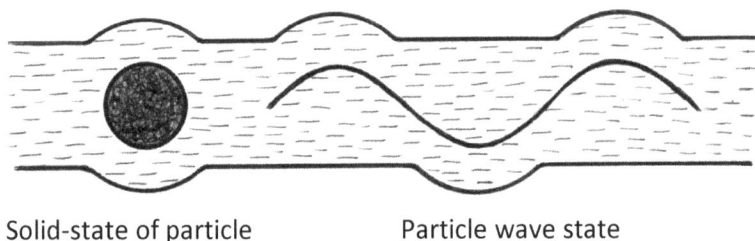

Solid-state of particle          Particle wave state

In this situation, the particle is situated in one micro-layer of space and is acting as an independent physical system.

The phenomena of entangled particles is happening when two particles are created simultaneously in one micro-dimension or on the boundary between two micro-layers of space dimensions. For convenience, we will consider only the wave function of the particles.

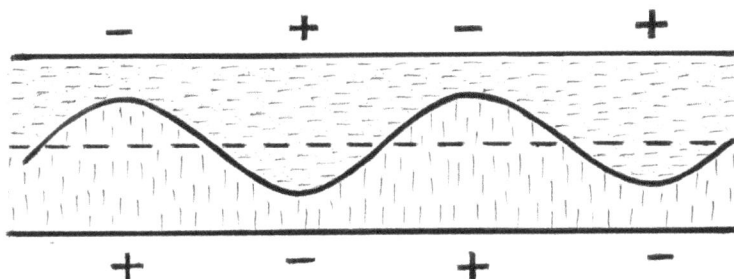

What are we seeing in this diagram? - The harmonious distortion of the boundary of space is creating two particles – one above the boundary, and one below the boundary. It is obvious that the two entangled particles must be in the opposite phase and change on one side of the boundary in instant affecting the state of the other side!

This is the reason why we always have two entangled particles, but not three, four, five, or more. The law of conservation of energy is dictating that any changes of the state (or phase) of the wave in one of the dimensions will affect the state of the wave in the neighboring dimension in instant. This is explaining how the "Information" between the entangled particles is traveling in an instant. - Because in the physical dependency between the layers of

42

space is not involved Time!  - The expansion of one layer into another changing the state of the second layer in instant!

Solving one by one the current "Mysteries" is a good indication that our concept is correct. This case is proving also that Space and Time are separate physical entities. - Because in the case of entanglement Time is not involved - this will be impossible if Space and Time are one substance.

This is just a simple explanation of the phenomena of particle entanglement and we should know that in reality the picture is a bit more complicated and there are involved all six dimensions and that the directional propagation of the waves is Time-related.  - (Refer to the graph page 61)


## UNDERSTANDING THE NUCLEAR STRUCTURE OF ATOMS, AND THE UNION OF QUANTUM MECHANICS AND NEWTONIAN PHYSICS

The range of unexplained "Puzzles" of Quantum mechanics is giving us an understanding that the micro-world behaving differently than the macro world. Currently, the "Standard Model" of Physics cannot provide us with a credible explanation of why the Law of Physics is different for the Micro and Macro-World. I have to assure you that the Law of Physics is the same for the entire Universe, and also is the same for the Micro and Macro world.

The assumption of the "Standard Model" that there are two different sets of Law for Micro and Macro world is a result of their incorrect fundamental model on which are constructed their theories. Their three-dimensional space configuration cannot explain the World which is based on six dimensions. The fatal flaw is that they are mixing together Space and Time in one impossible physical substance and this "mixture" is creating incredible chaos in their mathematical equations and is forcing them to create non-existence particles, nonexistence forces, and non-existing invisible substances.

The clue for the difference between Micro and Macro-world is coming from the fine structure of space and its influence on elementary particles and the fundamental forces. If we pay a little bit more attention to atomic structures, we will realize that the atomic structure of matter clearly is based on a six-dimensional spatial configuration and separation of the particles. That's why the stable electron shells of atoms are Six, the Quarks are also Six, their "colors" are Six, the Leptons are also Six. The particles are separated from each other by their spin (the spatial orientation). This rich variety of particles and orientations is possible only, if each particle is situated in a separate dimension, and just three spatial orientations are absolutely not enough to explain the incredible complexity of atomic structures. From this realization, we can draw the logical conclusion that on microscopic level Space is forming fine layers with a different orientation, which is providing the particles with their spatial identity and separation. - And exactly this is the difference between the Micro and Macro-world. - For the larger objects space is acting as a uniform physical medium, but for the Micro-world space has a fine composite structure. (See the graph below)

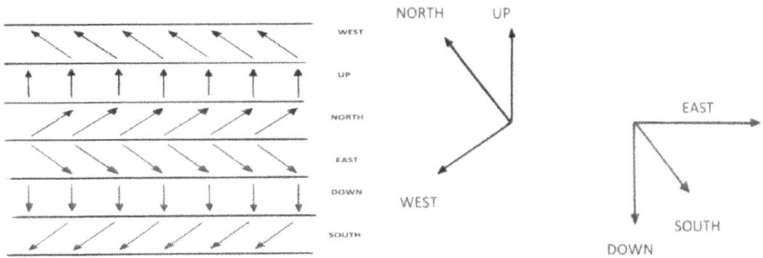

This is a graphical representation of the fine layers of space dimensions and the two pairs of opposing dimensions (on the right)

On a microscopic level, the spatial orientation of the micro-layers have a dominant influence on the particles, but as soon as the material structure grows above the thickness of the "layers" of Space Dimensions, Space starts acting as a uniform homogeneous space for the largest objects, and the Newtonian laws are taking over.

This explains the "weirdness" of Quantum Mechanics and why Quantum Mechanics cannot rule the biggest structures and is valid only for the microscopic world. – This is the currently missing unification point between Newtonian Physics and Quantum Mechanics.

When we apply the correct model to the atomic structure, becomes evident that we don't need a separate Law of Physics for the micro-world. – We don't need two separate Laws! - We need only a better understanding of the fundamental elements!

This is an example of how the six-dimensional space configuration is providing the many different angular momentums of particles. I will give you an example of how the six-dimensional configuration is providing a variety of space orientations: For example, the arrow (Down) can form <u>four</u> (three-dimensional) sets of orientation as the example next to it. - (Down, South, East); (Down, South, West); (Down, West North); Down, North, East). The same is valid for the North arrow. It also has <u>four</u> combinations with the other directions. This partially is explaining the weird numbers of the line of stability of the elements, because some angular momentums are incompatible with each other.

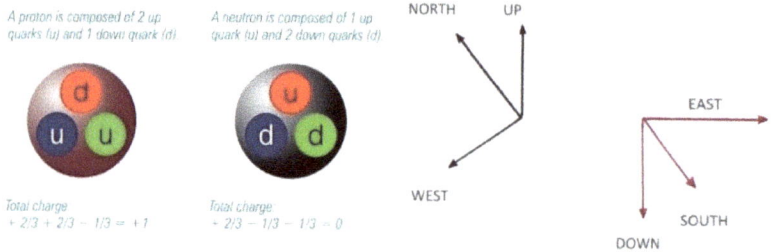

A proton is composed of 2 up quarks (u) and 1 down quark (d)

Total charge
+ 2/3 + 2/3 − 1/3 = +1

A neutron is composed of 1 up quark (u) and 2 down quarks (d)

Total charge
− 2/3 − 1/3 − 1/3 = 0

The Quarks are a good example of the three dimensional configuration of their internal forces.

The current "official model" of the atomic structure is unnecessarily complicated with the invented non-existing particles and forces. This is creating conflicting scenarios, and an unworkable configuration, where the atomic elements and forces are in direct conflict with each other.

To get rid of this mess we have to start from one simple fact – Matter and antimatter are annihilating each other when coming in contact. This simple fact is telling us that the atomic structure of matter and antimatter is held together by electromagnetic force because the only difference between matter and antimatter is their opposite electromagnetic polarity. The annihilation is no more or less a simple electrical ("Short Circuit"). This "Simple Fact" is telling us that we have to concentrate and understand what this simple fact is telling us! – That we have to concentrate to understand what is the configuration of the electromagnetic forces and polarities in the atomic structures to be able to hold all the elements together. If we do this, we will start from the correct position – from a solid fundament and can continue ahead and build a correct model of atomic structure!

In the official model, the attractive particles "Gluons" are in proportional dependency to the nuclear particles, and are evenly distributed (exchanged) between them. This configuration cannot explain the existence of mass deficit and mass excess of the nucleus. Gluons cannot explain how and why the mass deficit of the atomic nucleus is producing nuclear fusion. This configuration will allow the atomic nucleolus to be stable and to grow exponentially to any size, even to become bigger than the Universe, because the number of the "Gluons" is not restricted from the accumulation of more particles! – (This crucial fact obviously has been ignored).

When we start using logic, it will become obvious that the strong nuclear (attractive) force must be allocated only in the center of the atomic nucleolus. The attractive mechanism, which we have considered earlier, is working on the principle of energy cancellation between the objects. It is easy to comprehend that the opposing energy of the nuclear constituencies is canceling the space energy between them (in the center of the nucleus). The result is that the center of the nucleus becomes the point of partial cancellation of the space energy field - (as energy vacuum point), which acts as an attractive center for the nucleus. - This explains why the strong nuclear force has such a short-range - because is confined energy between nucleons and the center of the nucleus. That's why the strong nuclear force despite its enormous strength is not leaking out. This configuration is explaining also why there is a mass defect of the nucleus - (the missing binding energy). This explains why the atomic nucleus cannot grow exponentially and why become unstable beyond the 82$^{nd}$ element! The spontaneous nuclear decay of some isotopes of the lighter elements is a clear sign that the atomic nucleons are slowly gaining their missing "binding energy", which in effect is gaining of energy from the surrounding energy fields. And this small and gradual gain of the "missing nuclear energy" is clear proof of the energy cancellation nature of the attractive nuclear force, and that the actual source of the attractive nuclear forces, gravity and electromagnetism has the same origin and same attractive mechanism, which is a local cancellation of the space energy field! It is pointing also, that the "Weak nuclear force" does not exist!

We have to start paying attention to the fact that the structure of the atom is based on groups of six fundamental elements: - There are six stable electron shells. The proposed or "detected" six Quarks are named: Up, Down, Charm, Strange, Top, and Bottom. The known leptons are also six. The existence of six fundamental elementary particles with six specific "angular momentum" cannot be accidental, and this specific number is confirming the existence of six separate space dimensions, which are the actual instrument to separate those particles by giving them **six** different space orientations, which the top echelon of science is naming as "angular momentum" (or spin).

If the top scientists have known earlier, that their "angular momentum" is just ordinary six-dimensional space orientation, definitely, they have named the Quarks as East, West, North, South, Up, and Down. It is obvious that the variety of elementary particles can differentiate and be separated from each other by their different six-dimensional space orientations. It cannot be co-incidental also the fact that the stable electrons shells are **six**. It becomes obvious how the variety of elementary particles is differentiating and separating physically from each other because of their specific different **six**-dimensional space orientation, (or space dimensional separation). The neutrinos are the most abandoned particles in the Universe. They are also the smallest particles currently known. The physical properties of neutrinos are one of the most compelling pieces of evidence for the existence of **six** dimensions and **six**-dimensional space orientation because the neutrino number is also **six** – (electron; muon; tau, and their three antiparticles).

I quoted them as **six**; because the well-known fact is that the neutrinos and antineutrinos are not interacting with each other, and not annihilate each other. - Actually, they are the same particles, just having different (opposite) space orientations, which corresponding to the three pairs of oppositely oriented space dimensions - (east and west; north and south, and up and down)! The neutrinos are the lightest and the fastest known traveling particles. They are not interacting with anything, and do not annihilate each other. This physical property makes the neutrinos a good candidate for the first building block of matter. In such a case they could be instrumental for the nuclear forces too. The beta decay of quarks is pointing out that we don't have to ignore also the possibility that the quarks could be composed of neutrinos. The fact, that we cannot detect any internal quark structure is no evidence of lack of neutrinos there because we don't have any method for direct detection of neutrinos. The quarks have different weights and different physical properties. Four of the quarks are not stable. The different physical properties of quarks are telling us that they cannot be the prime building block of matter, but are composite particles with different quantities of energy, matter, and particles involved in their structure.

Modern science is sensing that the spatial structure cannot be uniform on a microscopic scale, but their concept of three dimensions mixed with the Time Dimension is stripping them of the ability to understand what is going on. Our understanding is that space is six-dimensional and the atomic structures are giving us clues of how the fine space layers are constructed.

The fine space layers orientation is putting the particles in different "Phases" not to interact with each other. - (Similar to the graph of three-phase AC current - page 69).

Atomic structure is similar to the layers of onion structure – the space layers are wrapped around the nucleus because space is an energy field and the created powerful energy "vacuum" in the center of the atomic nucleus inevitably will create a few spherical energy levels around the nuclear center. (see the graphs below) where the "cut off' projection is similar to the rings of a tree trunk. The different rings represent the different dimensions, and the distance between them provides room for the elementary particles to be in one, or be between two single space dimensions. This allows them to jump from one level to another, or to spread and be difficult to locate them.

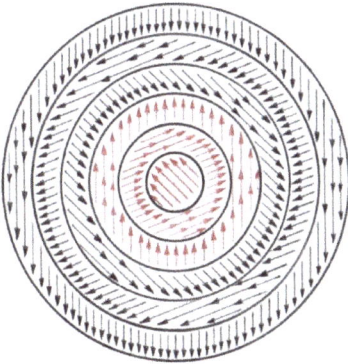

This is the fine space layers of the atomic structure

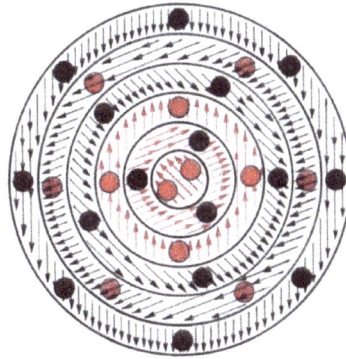

Shell configuration of atomic nucleus where the neutrons are separating the proton shells

The evidence for the shell configuration of the nuclear structure is coming from the fact that elements with up to three electron shells (the number of nucleons and electrons are related) having (nuclear mass deficit) - releasing energy in nuclear fusion. Elements with more than three electron shells having (nuclear mass excess) - or need additional energy to be able to fuse their atoms, (Because the next three shells are situated in dimensions having

48

opposite directional oriented energy field to the first three dimensions). We understand that confronting the angular momentum of particles will need additional confining energy. Our model explaining with ease the current "Enigma" of modern science - why when are we fusing together two atoms with a mass deficit they are producing much more energy than when we split two heavy atoms having mass excess?

Further proof for the shell configuration of the nucleus is the fact that when the nucleus is bombarded with electrons, some elements first emit neutrons, but some elements prefer to emit protons. This can be explained only with the shell configuration of the atomic nucleus, where the different shells are constructed respectively with an excess of neutrons or protons. The configuration of the nuclear forces is leading to the conclusion, that the protons are separated by neutrons where in the first, third, and fifth nuclear shells the protons are dominant, but in the second, fourth, and sixth shells, the neutrons have the majority. The currently unexplained variation of the nuclear forces in relation to the atomic number easily will be explained easily by the six-dimensional space configuration of the atomic structure. - The space energy cancellation is located in the center of the nucleus and the space energy deficit is just in the center of the atomic nucleus - (It is confined energy between the nucleons). This is the reason the strong nuclear force to have such a short-range and not to leak out. This configuration is explaining also how the elementary particles can be in superposition, and why the atoms are stable, why electrons are not losing their energy and are not falling and crashing in the nucleus. The six-dimensional configuration is explaining also how the different space orientation and space position of the elementary particles is separated and make them different from each other, but not stopping their physical interactions.

## SPONTANEOUS NUCLEAR DECAY AND WEAK NUCLEAR FORCE

From the chapters above, we reach an understanding that nuclear binding energy is an energy field cancellation between the nucleons. Also, the nuclear mass deficit is giving us an understanding, that the opposing energy of the nucleons is providing the space energy cancellation between them, (which is the strong nuclear force). This knowledge is giving us a good understanding of the structure of the atomic nucleus, where the energy cancellation is in the center of the nucleus. This nuclear structure making the outer shells of nucleons very weak attached, and further growth of the nucleus become impossible. In the heaviest elements beyond Led, we observe spontaneous nuclear decay. For an easy understanding of the concept of spontaneous nuclear decay, I will use the example from thermodynamics, where the individual atoms and molecules of matter are having slightly different energy levels from each other. The same phenomenon is valid for the nuclear energy level of different atoms. The overall decay rate for specific elements is uniform, but the life of a specific atom is dependent on the energy level of the atomic nucleons! - This is the real explanation for the reason for the spontaneous nuclear decay. - The outer shells of nucleons are losing at a very slow rate it's (excess) of binding energy, due to interactions with the surrounding energy fields. I will explain to you the reason - The elements with six shells, the angular orientation of the six dimensions is well balanced - their combined value is Zero! In the heaviest elements as Uranium, which have seventh electron shells there is included additional seventh dimension. The angular momentum of the seventh dimension is in direct conflict with the angular momentum of the first (the inner) dimension. - (It is the same spatial orientation). This is disrupting the neutral balance of spatial orientations and creates instability. The energy is confined, when is between two exact opposite strength and polarity points. The seventh dimension is destroying this balance, and the atomic force is not well confined anymore! This is the reason why the heavy elements with seventh shells are leaking radiation and decay. - (When the specific atom reaches the lower limit of binding energy, the atom decays). - It becomes apparent, that there is not involved any additional force or particles in the spontaneous nuclear decay.

The creation of the hypothetical "Weak nuclear force" is a result of the total misunderstanding of the nature of the attractive mechanism of the physical forces. Nobody ever has detected this invented "Weak" force. There is no single experimental observation of this hypothetical invented force. It is more than obvious, that the "Weak nuclear force" does not exist!

## SUPERPOSITION, MEASUREMENT, AND COLLAPSE OF PARTICLES WAVEFUNCTION

One of the biggest unresolved mysteries in Quantum Mechanic is the so-called phenomenon of "Particles Superposition". Quantum Mechanics states that the particles exist as a wave of probability in a state of superposition until observed. The act of measurement or (observation) is collapsing the wave function of particles and is forcing the particle waves to collapse and to form solid particles. Modern science has declared this to be a mystery, where the consciousness of the observer is collapsing the wave function of matter.

Based on this assumption, we are bombarded with baseless speculations that the world exists in a state of undetermined waves where nothing is real but is just a product of our minds. This strange and illogical physical phenomenon is giving the scientists also a reason to state that all the particles in the Universe exist in an undetermined state as a waves and the consciousness of the observer is collapsing the wave function of the particles and creating the material structure of the Universe. Even some of them are going much further and saying that we are living in a holographic Universe where nothing is real but is just a function of our imaginations.

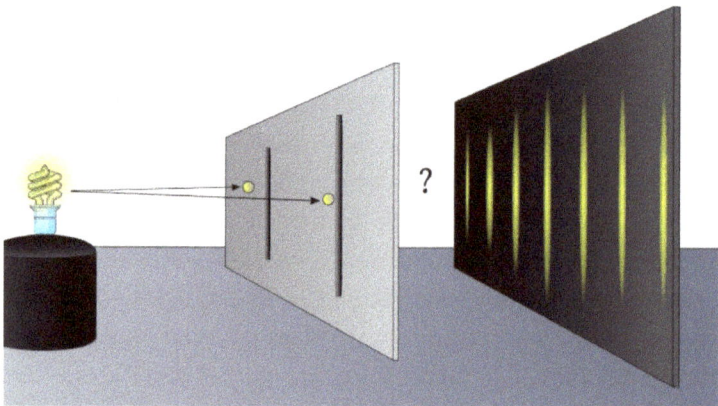

Double slit experiment

Not all scientists share this view and as a counterargument to this, they are asking questions: Before the observer was born who was collapsing the particles of matter? Where the observer is situated – in reality, or in imaginations? ... And is the observer real or he also is imagination?

Even Einstein has refused to accept the postulate of uncertainty. He insists that we are missing something important on a fundamental level. Even for an average person it is obvious that the state of uncertainty and probability is impossible to be the fundament of such fine-tuned Universe.

Before start consideration of this phenomenon I have to state that the mentioned above weirdness of the particles is real, it is well-study and well documented. The particle's behavior is weird, strange, and illogical. The strange particles' behavior is not connected and is not influenced by any other physical phenomena or function, which could give us a clue why this strange behavior is happening. Scientists even doubt that we ever will be able to find an explanation for it.

The problem is there when even a single photon or single electron is sent one by one at the time and if is not observed these single particles create a wave interference pattern in the screen behind the double slit. If any kind of measurement or observation is involved, the particle behaves as a solid particle. To all current physicists, this is the most mysterious, bizarre, illogical, and unexplainable phenomenon in the World. Everybody is scratching his head and asking – what is going on? In the last one hundred years, many thousands of clever versions of this experiment have been done. – "Delay choice" and "quantum eraser" versions, but the solution remains as distant and elusive as it was one hundred years ago.

TOE reveals that the fundamental difference between the Micro and Macro worlds is not that the Law of Physics is different for the micro and macro world. The difference is coming from the fine structure of space. - The biggest objects are crossing the boundaries of all six dimensions but the microscopic particles of matter are so tiny that they can exist just in one micro-layer of space. Each micro-layer of space is a single dimension which in some circumstances can act as a closed physical system.

Let's see how spatial dimensions are functioning. To understand better the concept of space dimensions, we have to use Time Dimension as an example to understand what the specific properties of a single spatial dimension are. We have to concentrate on Time's property and to realize that – <u>Time is spreading everywhere, and that Time is traveling in every possible direction because the direction of "Future" has no specific spatial orientation or direction.</u>

Absolutely the same scenario we will observe when we consider only one isolated single dimension. - <u>This Dimension will be spread everywhere and its orientation will be in all possible directions. Its orientation will not be determined.</u> You will ask why the spatial position and the orientation of a single dimension are everywhere and how and why its orientation is not determined?  The answer is very simple – In the unit called "Spacetime" each dimension has its specific orientation, but we are considering just a single isolated dimension. The single dimension definitely has orientation property. It is oriented, <u>but is oriented according to WHAT?</u> To have a defined orientation you must have a Reference Point! <u>Without a reference point the term "orientation", "position" and "direction" have no meaning!</u> – This is the answer to this question.

Understanding this we can come back to our question of particles superposition. To have a stable and clear interference pattern the scientists are isolating the experiment from the interference of the surrounding. By doing this they are creating a "Sterile Scientific Environment" for the experiment. The "Sterile Environment" is the core of the problems for the scientists because by isolating the experiment from the interference of the surrounding environment they are creating an "Isolated Closed Physical System". - Exactly this is the reason for the so-called "Mystery" and strange particles' behavior. Here is the explanation of what happening in this closed physical system:

The particles are small enough and are able to fit and exist inside the boundary of single dimensions. We just reach an understanding that a single dimension without a reference point is everywhere - (Same as Time). The position and the orientation of a single dimension and the particles in it are undetermined until a reference point applies! - And this is the key for understand why when is unobserved the particles in the experiment are everywhere. – Because there is no reference point! <u>The reference point is the physical mechanism of providing location!</u>

The particle in the boundary of a single dimension we can consider as to be inserted into a closed physical system. To understand the concept better, I will give an example: - A single passenger is traveling in a bus with fully closed curtains. (The bus represents the Single Dimension and the passenger represents the particle). The bus curtains are closed and the passenger inside cannot see in which direction the bus is traveling – forward, reverse, North, or South. Outside observers also cannot know where exactly the passenger is sitting – (his position). – This exactly is the typical situation with a closed physical system – the physical state, position, and direction are undetermined when is no passing information in or out of this system. Lifting the curtains of the bus is changing dramatically the situation. Suddenly, the passenger can see the direction of traveling and the outside observer can see exactly where the passenger is sitting (his position). The act of lifting the bus curtains is representing the act of "observation" in the double-slit experiment. - It is letting the information get in and out of the closed physical system. – It is an act of opening the closed physical system which is providing the information of direction and position. In short term, it is providing a reference point. The observation is not creating reality! – It will be a similar situation, when the bus curtains going up and we can see the passenger to claim that we have created the passenger.

The act of measurement (or observation) in the double-slit experiment is providing a reference point. We have to realize that this is just a simple explanation. We have to understand that not only the conscious knowledge of the observer provides the reference point, but many other factors can do the same. Even if a single photon pass true the undetermined cloud of probability of particles or atoms, this single-photon will do the trick – it will provide the reference point and the undetermined particle suddenly will appear. Soonest there is a reference point, the single dimension will have a defined specific orientation and the particle in it also will have its precise position!

In the end, we have to realize that this is no more or less a physical mechanism of providing a reference point to an isolated micro-physical system! - There is no mystery involved and also there is no such thing as an observer, who magically creates reality and material structures of the Universe.

The conclusion is that the particles can exist in (superposition), but normally they are just particles. The result of the experiment doesn't mean that the observer is creating reality rather that consciousness and information are active part of the physical organization of the Universe. The transformation between the wave function of particles and the solid state of particles is a very complicated physical process that is not well understood yet.

One of the pieces of evidence for the correctness of this explanation is the fac
that in an experiment you can choose the direction of looking into a pair of
entangled particles. In this way, prior to the experiment, you are determining
the direction of the single dimension where the entangled particles will be
found. When you chose to measure entangled particles you can choose any
direction – horizontal, vertical, or any other position. The entangled pair will
fit in your chosen direction because you are providing the reference point!
There will be only one condition – the particles always will be oriented
opposite to each other! The image below is a graphical representation of a
wave function of entangled particles in a single layer of spatial dimension.

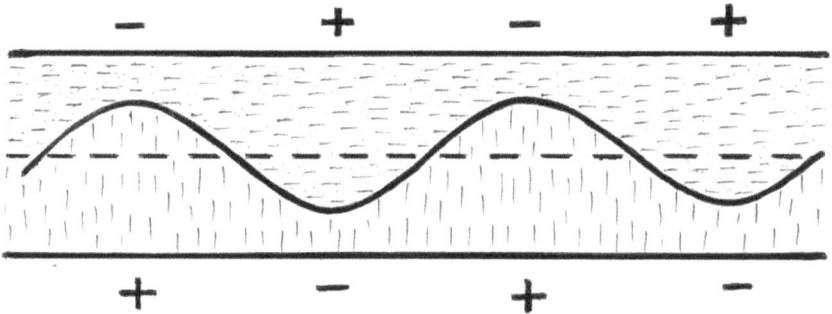

The sinusoid is dividing the pair. One is above the sinusoid; the other is below
the sinusoid.

It is obvious that the entangled pair of particles can exist only in the opposite
phase! Any changes in the state of one particle - (one side of sinusoid) lead to
instant change to the other particle. This is obvious and easy to understand
and there are no mysteries, no puzzles. Just pure logic in the boundary of the
Law of Physics!

## PRIMORDIAL SPACE; ORIGIN OF THE SIX-DIMENSIONAL SPACE AND THEIR ENERGY FIELDS

From the chapters above, we reach an understanding that the origin of all
energy of our Universe is coming from the disturbances of the equilibrium of
the six-dimensional space. This realization is bringing us close to the
primordial source of the Universe… But what will be this single and ultimate
source of everything around us? Do we have enough information to find it
out? It is obvious that our Space is not uniform, but is a composite structure.

Every composite structure contains different or the same materials which usually have a different structural orientation, which is giving new and unique properties to the composite material. In our case, Space is formed from six identical to each other Space Dimensions. This leading to the realization, that the prime source of Space should be a substance similar to a single dimension. I will give you an example of what exactly that means. - We also producing composite structures of a single material - for example, the oldest and simplest composite structure we have produced is plywood. The source material is wood, but the wood is not a uniform material and when we are slicing the wood into tiny layers, and glue them together we are making sure that the <u>direction of the fiber on each layer is on a different angle to the next one</u>. By doing this, we are producing material with different properties. The structure of our space looks very similar to the structure of plywood. – It contains <u>fine layers of the same material having different structural orientations!</u> It is logical that just one single dimension can be used as a primordial source of our space.

Let consider what physical property will have an isolated single dimension – If we are situated just in one dimension, there will not be Time, there will not be Matter, there will not be Energy (Physical energy) - the energy which we know. The fact that a single dimension has its own energy field, specific spatial orientation, and directional propagation will mean NOTHING! The energy we know is a disturbance of the equilibrium of space, but in a single dimension its energy field is uniform and there is not any source of disturbance! The spatial orientation and directional propagation are directional, but directional compare to WHAT? We have to realize, that in the scenario of a Single Primordial Dimension the terms -Time, Energy, Direction, Orientation, Propagation <u>are losing their meaning!</u> The state of a single dimension we can describe as "**Phenomena**" without active Physical property! - This explanation is completely different than the explanation of mainstream science, which has proposed "**NOTHING**" as a primordial source of the Universe! The existence of the primordial Dimension is not Nothing! - The Single primordial Dimension is not possessing Physical properties in the way we know them, but at the beginning, we started our considerations with a clear understanding that there exist also non-physical properties, which are real, but they are not Physical. The same scenario we have with <u>the Primordial single dimension! It is a real phenomenon, but is not Physical and definitely is not NOTHING!</u>

The Physical properties in the Universe are coming as a result of the disturbance of the equilibrium of the six Space Dimensions. - The energy source of our Universe is coming from the different orientations of fine layers of Space, and the manifestation of the energy is a disturbance of the boundaries and the equilibrium of the layers of our six-dimensional Space! - Everything else could exist, but won't be physical!

I believe that we have reached the limit and the boundary of our current knowledge and we have succeeded to define the primordial source of Space. For the origin of Time I cannot comment, because the Time dimension is a uniform energy field, or the orientation of the Time energy field is parallel to its directional propagation. To be determined the exact parameters of the Time dimension is necessary serious study and experiments, which current science is not ready to do. I am glad that we have succeeded to determine the primordial source of the Universe which has just three basic components – Consciousness, Time, and the Primordial Dimension. The Universe is built on the principle of opposing symmetry. That means in the creation of our Universe must be created also another Anti-Time dimension, which will cannot be left unused, but will be used in creation of a parallel Universe.

We have to stop here because to continue further will be just pure speculation based on fantasy but not on facts. It is not good to do science-based on speculations.

Still, we are left with the open question: Where the Consciousness come from and is the Primordial Dimension always existed?

## UNION OF ELECTROMAGNETISM, STRONG NUCLEAR FORCE AND GRAVITY

In this chapter, we will consider Electromagnetism and Strong Nuclear Force (SNF) and their relation to Gravity. For Gravity, we will allocate a separate chapter.

We need to find the origin of these forces because the understanding of their origin will help us to understand their difference and their similarity.

Lately, we will find that apart from the difference of their polarity, the electromagnetic and strong nuclear force virtually is the branches of the same force and we can call them - "Electro-Strong Force." After that, when we find that Gravity belongs to their Group, the name will grow to "Gravity- Electro-

Strong Force." Probably, such a complicated name is not practical and the three forces will preserve their original names as separate Physical forces belonging to the same Group. First, we have to find the reason why I am making this bizarre claim?

Don't be alarmed my friends, I know that the claim that the origin of these three fundamental forces is the same is very hard to digest, but believe me, like everything else in Universe even this will be simple and very easy to comprehend. - We just need to use a correct fundamental configuration to put our forces there, and to use logic to consider their interactions! Electromagnetism is based on polarization and specific spatial orientation. Our understanding of the positive and negative electromagnetic polarity is coming from the understanding that Space is not Three-dimensional, but is Six-Dimensional. The existence of two sets of three oppositely oriented space dimensions - (refer to the diagram below) - ("B" and "C") is shading light on the origin of the positive and negative charges. This configuration is explaining why Protons and Electrons are not annihilating each other and why electromagnetism has repulsive and attractive properties.

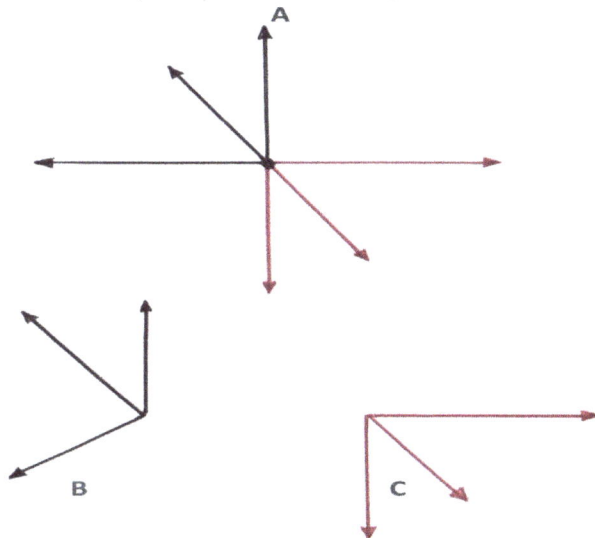

This is the six-dimensional Space configuration which contains two sets of opposing dimensions, which producing the opposite electromagnetic polarity

We have to understand that either - the negative and the positively charged particles having positive energy, and the crucial difference between them is coming from the opposite space orientation of their set of spatial dimensions where the charged particles are created and operate. (In Physics the spatial orientation of particles is called Spin).

To make it easy to understand the concept of polarization I have to explain that the spatial orientation of energy fields or particles is making them to act as directionally orientated vectors or forces.

In the diagram "A" above is a representation of the six vectors of Space. For easy understanding, I will name them as the six spatial directions - East, West, North, South, Up, and Down. It is evident that North, West, and Up (the black vectors) having opposite directions to East, South, and Down (the red vectors) The opposing space orientation of the two sets of vectors is producing the opposing electromagnetic polarity of the particles - (See the diagrams above) I would like to explain that this graphical representation is just a simple visual understanding of the concept of how the opposite spatial orientation is the mechanism, which provides positive and negative charges. In reality, space configuration is much more complicated, or I can say that it is extremely sophisticated. We just have to know that each spatial dimension is operating everywhere and is at each point of space. There also is involved the element of Time. We will keep it simple for an easy understanding of the fundamental principles of the physical system of the Universe.

I will provide an example of how the orientation of spatial directions is forming and shaping the electromagnetic force:

From the chapter "What the Energy is" we learn that energy is the disturbance of the balance of the space equilibrium. That means that the energy in one dimension will push and expand its boundary into the next dimensions – (see the diagram "A." below). The tendency of the system to restore the equilibrium is creating the response of the disturbed dimension. This response is dynamic and is space-oriented. In the diagram, "B" is visible how the current with Up direction is causing the formation of an oriented magnetic field in the neighboring dimensions. The result is the well-known - the "Law of the Right Hand," where the tam is pointing in the direction of the electric current and the fingers are in the direction of the magnetic field. (see the graph below)

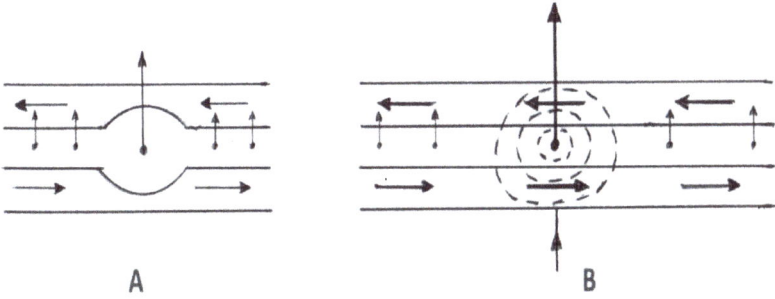

A                                    B

(**A**) The passing current is distorting space boundaries.
(B) The disturbance creates an oriented magnetic field.
(See also the graph below)

It becomes evident that the orientation of the neighboring dimension is providing the direction of the magnetic field caused by the current. For better understanding, I am providing the graph below, where become easy to see that the propagating electric field (in the middle dimension) is disturbing the energy equilibrium of the neighboring (Top and bottom) dimensions, and as a result of the energy disturbance, there is created magnetic field which follows the orientation of the housing dimensions.

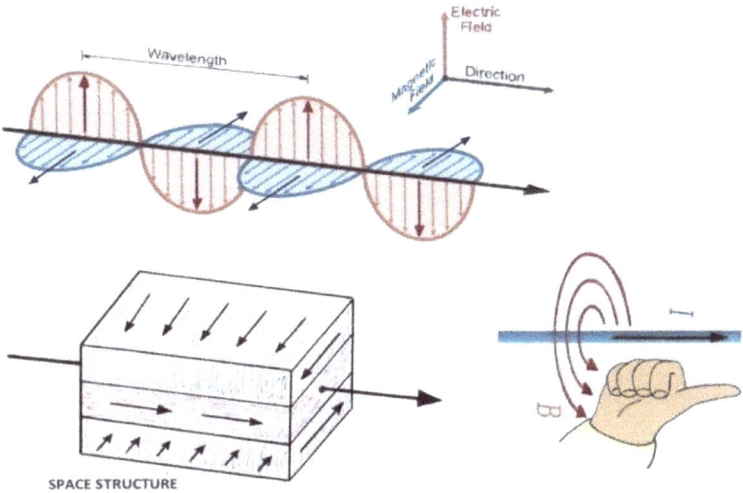

SPACE STRUCTURE

The graph on top (The sinusoid) is the current graph from the books of electromagnetism. There is visible, that the three-dimensional vectors are not representing the real configuration of the wave function of the electromagnetic wave. To be correct, this graph always should be

60

represented as the graph below, because the explanation of electromagnetism needs more than three dimensions to represent reality! - Notice, that the electric and magnetic fields in the diagram above are represented only with one single vector, but in reality, they are spreading also on the other side of the directional vector. The current science restricts itself in the boundary of three dimensions and is not able to produce even a realistic diagram of propagating electromagnetic waves.

The graph from above always should be represented as the graph below, because the explanation of electromagnetism needs a six-dimensional orientation to explain correctly the dynamic relation between electric and magnetic fields. In this graph are involved five vectors – (directions).

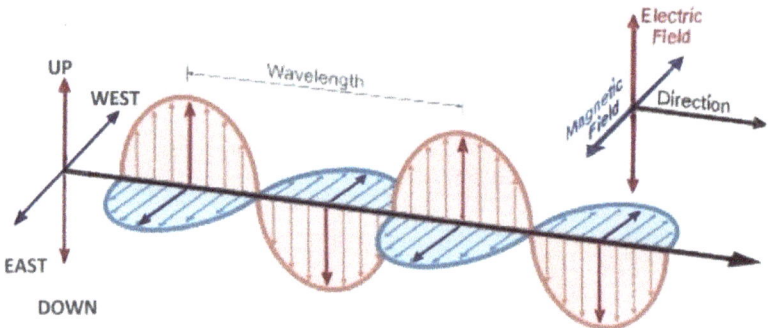

The directions of the current and magnetic field are a result of the orientation of the space dimensions - the (Right Hand Law)
We have enough dimensions to represent the direction of forces

From the previous diagram, become evident, that the electric wave traveling in the middle (red) dimension will push the boundary of this dimension into the two neighboring dimensions – (below and above). The disturbance of the space equilibrium in the two neighboring dimensions will generate a harmonic oscillating magnetic field, which will have the orientation of the dimension where this magnetic field is generated. – This orientation of the magnetic field will create a circle of an oriented magnetic field around the axis of the current - The "Right-hand law).

61

It will be necessary to understand that from each reference point of space, the space directions are propagating outwards (see the graph on page 22), and this is producing the phenomenon that the El. Magnetic current can run in any direction of Space, and the resulting magnetic forces always will line up in the same way, no matter in which direction the primary force is directed! The six-directional space is more complex than the currently assumed three-dimensional configuration of space. These complexities of Space are providing the incredible dynamic and versatility of electromagnetism.

For example, the carrier of the electromagnetic current – the electron has to rotate 720 degrees to come back in its original position. This is hard to comprehend but is a fact, which we have to admit. We have to realize that three-dimensional space has $360^0$ degrees only! - But six-dimensional space configuration have720 degrees and the electrons are clearly showing us what the reality is! It needs a bit more explanation - the easy way to understand the concept of $720^0$ (six-dimensional) rotation is based on two perpendicular axes of rotation. - One axis in a horizontal plane ($360^0$) and one more in a vertical plane ($360^0$). When you rotate only $360^0$ degrees the system is ending upside down, because the particle is rotating simultaneously on two perpendicular axis - $90^0$ on vertical and $90^0$ on horizontal axis = $360^0$. To restore the original space position, you have to rotate it another $360^0$. I have to remain you that this phenomenon is valid only for the micro-world of Quantum Mechanics. For the macro-world, space is acting as a homogeneous substance and we are not experiencing such weird effects. This fact alone is excellent proof for the correctness of my explanation of the existence of fine layers of the six space dimensions. (See the graphs below)

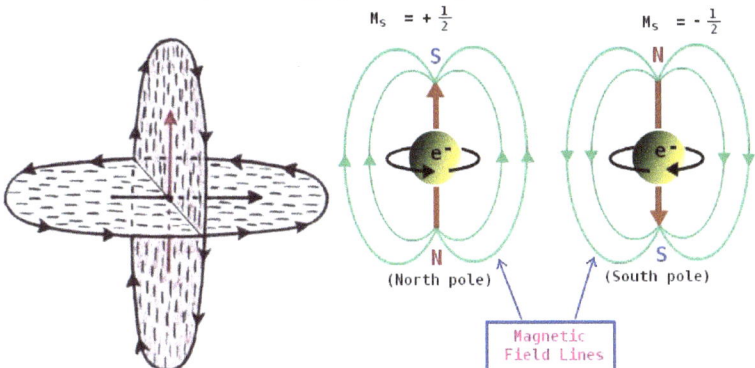

$M_s = + \frac{1}{2}$

$M_s = - \frac{1}{2}$

(North pole)

(South pole)

Magnetic
Field Lines

360$_0$-degree rotation is changing electron polarity and orientation.
720$^0$-degree rotation is returning the electron to its original position
because the particle is rotating simultaneously on two perpendicular axis

We don't need complicated mathematical formulas to understand how the direction of spatial dimensions is forming the direction of the two related forces (electric and magnetic) and where the law of the "Right Hand" is coming from. From the graph above become evident that the traveling electric wave in one layer of space is disturbing the energy equilibrium in the neighboring dimensions. The disturbance of the "energy pressure" or (the equilibrium) of the dimensions is creating the magnetic field which is oriented parallel to the orientation of the disturbing dimension. - This is how the orientation of the space dimensions is providing the direction of the magnetic field around the electric current!

For simplicity, we are considering the electron as a solid object but will be good to know that the picture is much more complicated, because the particles also exist in form of waves, and the wave functions have a few components as spatial orientation, wavelength, frequency, energy level and direction of propagations.

Modern science knows that the elementary particles are just standing waves around a negative center. Despite this knowledge, modern science claims that some sort of "solid" Quarks is spinning inside protons with speed close to the speed of light and their "kinetic" energy is 99% of the missing mass of the protons. This speculation cannot be further from the truth! The Strong Nuclear Force of Protons and Neutrons are not the kinetic energy of the Quarks! We have to understand that <u>Kinetic energy is not providing attraction</u>! - This is nonsense and another absurd and baseless assumption of modern science!

Strong Nuclear Force is a fundamental attractive force, which is based on the attractive mechanism explained in the previous chapter. The Strong Nuclear Force is responsible for 99% of the mass of the protons and neutrons, and as a consequence of this, is also 99% of all the mass of the atoms and the solid objects. If we ignore the remaining 1% we can say that everything we can see around us (and even ourselves) - is just a form of strong nuclear force.

We cannot continue forward to do science if we don't understand the origin of the 99% of matter.

We will continue by coming back to the annihilation of matter and antimatter to see what we can learn from this phenomenon because this process is giving us a revealing clue for the origin of the SNF. If inside the nucleus exist a different form of the nuclear bond rather than electromagnetic, then the process of annihilation will not work, because the SNF has superior strength in comparison to the electromagnetic. The annihilation is destroying completely all internal nucleus structures and is turning them into pure energy! We have to realize that <u>the electromagnetic force will never be able to destroy the atomic nucleus, if SNF is not also an electromagnetic force</u>, because SNF is about 137 times stronger than electromagnetic force! Such complete disintegration is possible only if the strong nuclear force also is an electromagnetic force!

If SNF has different physical properties, electromagnetism will not have any destructive effect on the structures hold by the superior nuclear force. This is the fact, which is giving us a new and fundamental understanding that <u>Strong Nuclear force has combined "Zero" polarization and is acting as non-polarized, but still is an electromagnetic force</u> that is operating in all six dimensions!

The existence of the mass deficit and mass excess of the atomic nucleus is evidence of extraordinary significance! The Standard Model has no credible explanation of the mass deficit and even is providing a wrong graph, where the mass deficit has a positive value. (See the graph below). We have to realize, that **mass-deficit** as nuclear bond means only one thing! - <u>Energy deficit is the attractive bond in the center of the nucleus</u>! This exactly is the concept of the Universal Attraction reviled in a previous chapter. The nuclear mass deficit and nuclear mass excess are direct evidence in support of the concept of the six-dimensional origin of nuclear attraction. The atomic mass deficit and mass excess are revealing clues for the understanding of SNF. We know that each particle is a standing wave around the energy deficit center. What happens, when the protons and neutron is combining and are forming a nucleus? - They are combining their energy deficits into a common one. From chemists, we know that two atoms can combine and share one electron. From this example become evident that when proton and neutron are combining together in a nucleus they are sharing one common energy-deficit center. In the first three layers, the spatial orientation is similar and they adding their strength to each other. This is creating a reduced energy demand, which is manifesting as an "Energy mass deficit." But the orientation of the next three

layers of space dimensions is in direct confrontation with the first three. This i
increasing the energy demand, which we seeing as "Mass excess."- This is the
reason for the mass deficit in the elements up to Iron and mass excess in the
heavier than Iron elements.

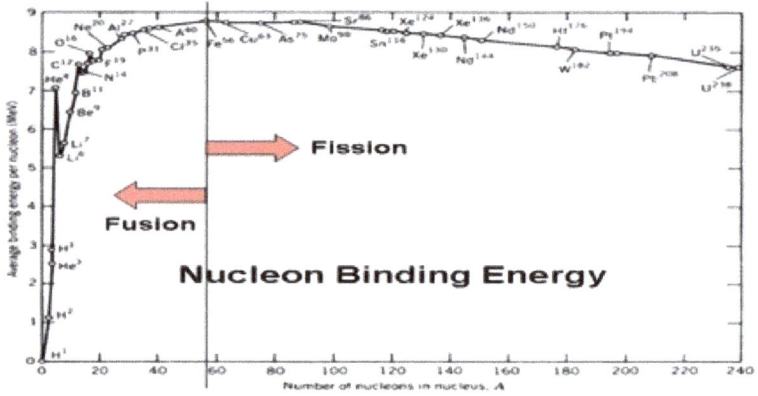

Nuclear mass-deficit having the elements with up to three
electron shells. The fourth electron shell is where starting
the nuclear mass excess, or the (energy excess) - Iron.

**B.**

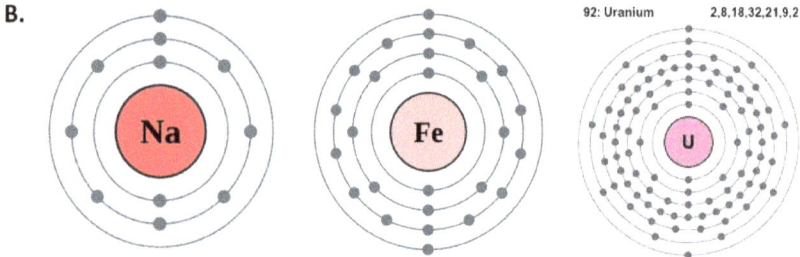

The electron shells are a good indication of the spatial dimensions

In the graph on the next page is a graphical explanation of this. - The energy-
deficit center formed from similar oriented forces should be smaller than the
sum of their constituencies before the combination. - And exactly this
"reduced energy demand" configuration is the manifestation of the missing
mass of the nucleus in the first elements! With the increasing numbers of
shells in the post-iron elements, the repulsive force of the opposing
dimensions in combination with the increasing distance from the center has to

be provided with more binding energy, which the nuclear particles do not have. It becomes evident, that the elements with a bigger number will need <u>additional externally added energy</u> to counterbalance the repulsive energy of the increased number of protons. - This externally added energy is the manifestation of the so-called "Mass Excess" of the Nucleus.

(See the graphs above and below)

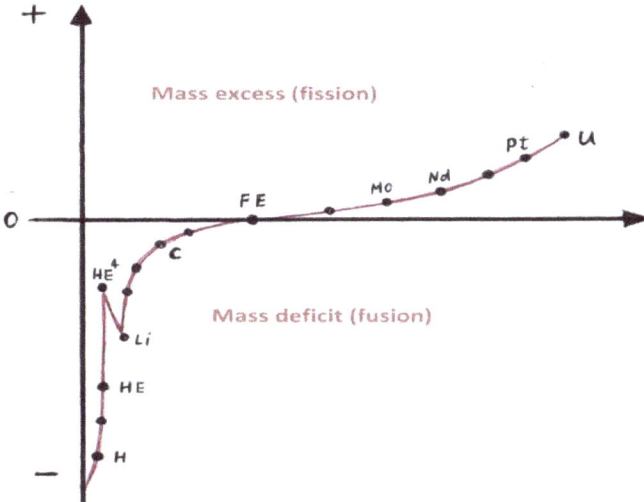

The correct graphical representation should look like the graph above, where Iron has (zero) nuclear mass deficit, (or mass excess).

From physics, we know that we can represent forces with different directions as vectors and their resulting force. In the nucleus, the different orientation of the space directions is producing resulting forces with different directions. On the graph below we can see how the different orientation of forces require different strength or (size of the resulting SNF) to be able to hold together the nucleus. - Where the diagonal is the resulting (Strong nuclear force)

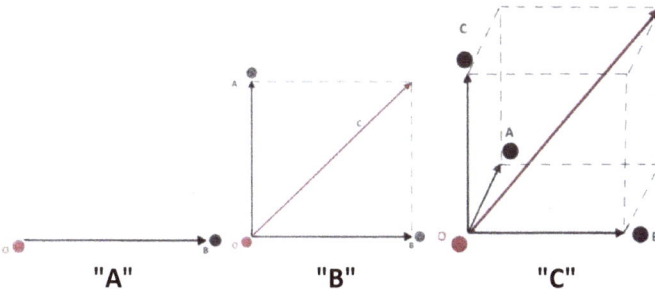

"A"            "B"            "C"

Graph "A" is representing the SNF in a Hydrogen atom (Proton - electron)

The size of the resulting (diagonal) in the graph above "B" and "C" is the graphical representation of the size or of the strength of the Strong Nuclear Force which is necessary to hold together a more complicated atomic nucleus - the different orientation of forces in the nucleus of the heaviest elements need bigger binding energy!

When we compare the size of the resulting force vectors - ( the diagonals) - in graphic "A", "B", and "C" wherein "A" is a single-dimensional orientation (Hydrogen). In "B" (two-dimensional orientation) the vector is 30% longer, but in "C" (three-dimensional orientation) the diagonal vector (SNF) is double than "A" – This is a graphical representation of the necessary strength of the strong nuclear force in the atomic nucleus. It is evident that the more complicated nucleus will need more energy to hold together the nucleons than the simple ones. This is explaining where the mass deficit and mass excess are coming from. With the increasing number of elements, in their nucleus, the number of protons is increasing and this also is increasing the repulsive force between them. The increased repulsive force will require a stronger nuclear force. We know that increasing the energy content of a closed configuration of a physical structure is increasing its mass! - This is exactly what we are observing in the atomic nucleus of the elements with a higher number - gradually and proportionally increase of the binding energy and the mass of the atomic nucleus, which we are calling "Mass-excess".

We have found the reason for the so-called mass deficit, and mass excess of the atomic nucleus. Still, we have to find out how the by-polar attractive electromagnetic forces are configured in order to be created uniform attractive SNF without magnetic property and polarity which is able to attract positive and neutral particles.

From the chapter above, we reach an understanding that electromagnetic force has positive or negative polarity, depending on the set of (three)dimensions where they are operating - (See the graph page 59), It is easy to understand, that the force in order to have neutral electromagnetic polarity, the force must operate equally in all six dimensions. This will balance the magnetic and electrical polarity to Zero. As a result of this balanced polarity, we are observing that SNF is attracting with equal strength the positive, negative, and neutral particles.

I have included the (negative) particles just to explain that SNF will be absolutely the same - (having neutral electromagnetic polarity) in the nucleus of antimatter.

I believe that all these facts are giving us the confidence to formulate correctly the physical property of the Strong Nuclear Force. - The Strong Nuclear Force is an energy deficit operating equally in all six dimensions where its combined polarity is zero! - That's why the Strong Nuclear Force does not have polarity! - This is the secret of how the SNF is able to attract - the positive and neutral particles of matter. To understand better the lack of polarity and how the energy deficit of (SNF) works I am providing the graph below:

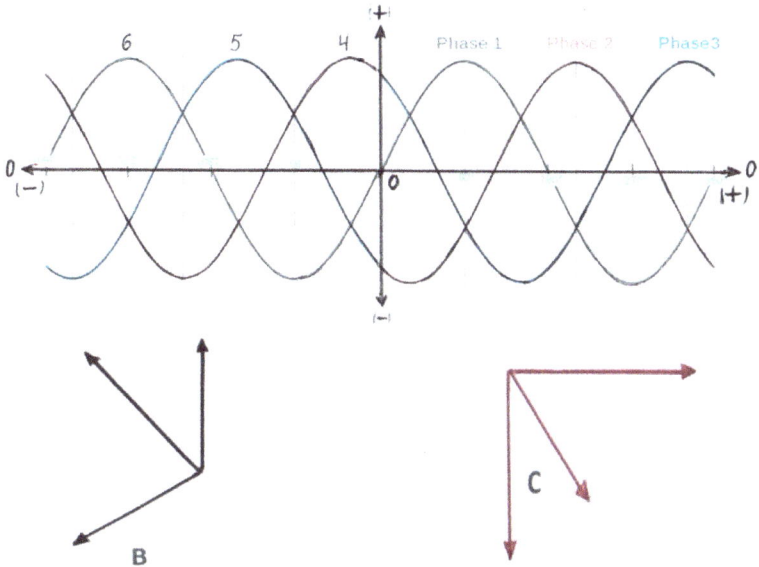

In this diagram, ( the sinusoids) on the left and on the right of the vertical axis are representing the particles with different polarity, (opposite space orientation "B" and "C").

The horizontal axis "0" is representing the SNF which is having (neutral polarity) but is attractive to both - positive and negative charges. (Same, how the earth is neutral to the three phases of three-phase electricity)!

The six-dimensional SNF configuration can be represented as the graph above of three-phase electricity, but with two opposite directions - (minus to the left and plus to the right)

The nucleons are separated by their six-dimensional orientation. The difference is that the first three dimensions going right of the vertical axis and producing positive polarity, and the next three dimensions are running in the opposite direction - on the left of the vertical axis and producing negative

68

polarity. The Strong Nuclear Force (SNF) representing the horizontal axis "0" (Zero). The SNF can have any potential, but to the particles included in the six dimensions, the SNF always will have an Attractive Value! The SNF is operating simultaneously in all six dimensions and has combine zero space orientation, which is manifesting as combined zero polarity. Electromagnetism is based on the opposing polarity of the two sets of (three opposing each other dimensions). With this understanding safely we can say that SNF and Electromagnetism have the same origin and we can unify them and call them "Electro-Strong Force."

I will provide one important fact in support of this conclusion: We know that gravity always is proportional to mass. Despite this, the Neutron Stars have 3,000 times stronger Gravity than should be according to their mass. It becomes obvious, that the strong electromagnetic field of Neutron Stars is producing a substantial excess of Gravity. The same thing we are observing with Quasars. - Their strong electromagnetic field is mimicking enormous mass, but in reality, they are not such massive objects of how mainstream science suggests. - This is explaining also where the gravitational bond of the Galaxy is coming from, and why their stars are not flying away! This is the facts that explaining that in the Universe exists only one fundamental force - The Strong Nuclear Force.

Gradually we reaching an understanding that Gravity can be produced by electromagnetism and SNF, but SNF is a complete form of electromagnetism. In the next chapter, we will consider the origin of Gravity and the relations between Gravity, SNF, and Electromagnetism.

I am happy that we were able to resolve the long-standing "Puzzle" of how the Strong Nuclear Force is able to attract magnetically neutral particles.


## ORIGIN OF GRAVITY AND ITS RELATION TO ELECTROMAGNETISM

Before start, our consideration of how our model is explaining the phenomenon of Gravity will be good first to get familiar with the strange and weird manifestation of the Gravitational force and how the Theory of Relativity explaining this phenomenon.

The main reason for the acceptance of the Theory of Relativity is coming from its claim that the curvature of space is "eliminating" the need for the emission of gravitational force from the bodies of matter. The main problem with the nature of Gravity is coming from the fact that all celestial bodies are emitting Gravitational force continuously for billions of years. Such force emission can only be at the expense of the internal energy storage of the atoms of matter. Such emission inevitably will deplete the nuclear bond of the elements, and they will start to decay gradually and continuously until they disintegrate in their original prime form of energy. The fact that the elements are stable and exist for billions of years unchanged means that the matter is not emitting Gravitational force!

The proposed of Einstein curvature of space looks like is solving this problem, but in close details observation of the claims of the theory, we can see a very significant discrepancy and anomalies:

In order for the space to be curved near massive bodies, Einstein also assumes that these bodies are emitting force that is bending the space. Soon Einstein manages to bend "his space," he just "forgets" for the existence of the force emitted by the massive bodies. Why? Wherefrom is coming, and where is going after this force? And In the end, Einstein is not eliminating the existence of the emission of gravitational force at all! - (He just ignoring it)!

Unfortunately, you cannot prove anything by ignoring the facts! – First, Einstein postulates that Gravity force does not exist, and after that, Einstein starts making predictions for how the material bodies will behave in a strong gravitational field! Even scientists have "discovered" the "Frame Dragging" effect, which the celestial bodies are inserting into the surrounding space!

I am sorry, but mainstream science has to come clear on which principles are based their "Standard Model." They are rejecting the existence of Gravity but are using the gravitational field. They reject that space is a medium, but accepting that the "Gravitational field is dragging space." You cannot reject and use the same phenomena at the same time! - This is a self-contradiction on a fundamental level! Unfortunately, the problems are not ending there! There are a big question and discrepancy with the mechanism of how the curvature of space is claiming that it is "producing" the attractive gravitational effect. - Here is one example of the concept's explanation (the image below) – They are using the Earth's gravity to make the boll rotate around the center.

I am suggesting this experiment be done in the International Space Station, where is no Earth's gravity to see what will happen with their "space curvature attractive" mechanism?

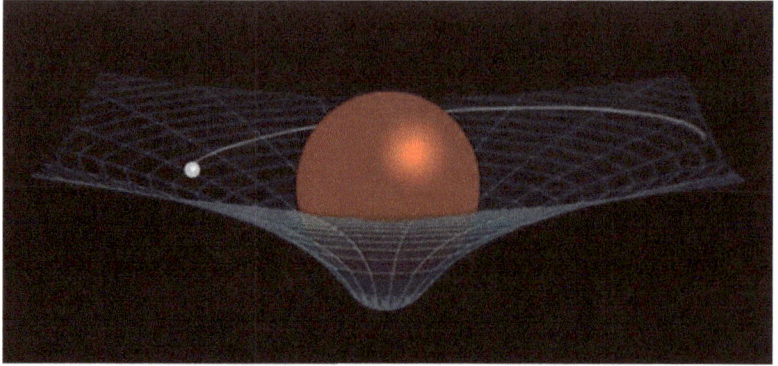

It is obvious that this explanation cannot work without the Earth's gravity from below.

Even if we close our eyes and accept the earth's gravity in their experiment, the" space curvature" somehow works for moving objects, but cannot explain why stationary objects should move toward the center? Why two stationary objects are attracting and moving toward each other? - This is a fundamental question! -We know that space is not inserting force on the objects; even Theory of Relativity is not claiming this because if it is claiming this, it has to explain where such force will come from? Still, they are avoiding to explain why two stationary objects in space attract each other? If a theory of such significance cannot answer elementary questions like this, what kind of theory is it?

Still, this theory is not explaining where the force of attraction (Gravity) is coming from and the fundamental question of what and how is producing attraction? - The claim of curvature of space is not explaining why the "curvature" of space is affecting in different ways objects traveling at different speeds? - Because the "curvature" of space must be the "straight line" for all of the traveling objects in this curved space regardless of their speed! The fact that the objects have a tendency to preserve the strait line of traveling is proving that these objects are situated in space with a straight configuration (not in curved), and the Gravity Force is attracting and is bending the trajectory of the traveling objects more or less dependent of their speed - (This is the simple and well-known "ballistic effect") –

(See the images below)

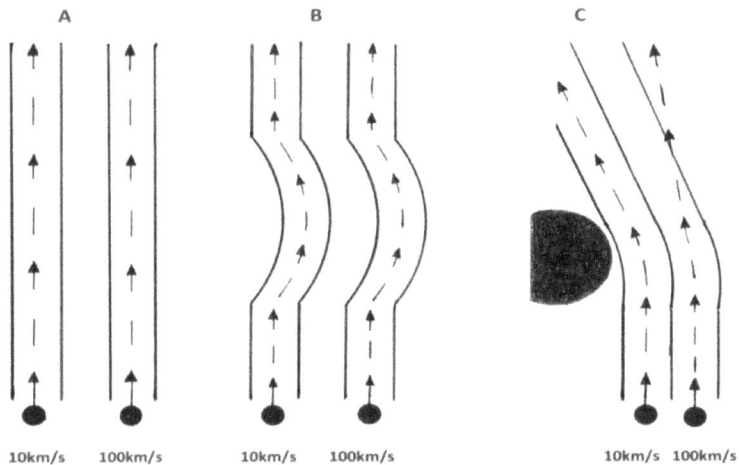

|   A   |   |   B   |   |   C   |
| 10km/s | 100km/s | 10km/s | 100km/s | 10km/s | 100km/s |

(Flat space)         (Curved space)              (Reality)

The traveling objects strictly must follow the shape of the space
where they are situated regardless of their speed... but
what happens in the diagram "C" and why the
the fast-moving object is flying away?

- In the left diagram -" A" is the path of two objects traveling at
  different speeds in our "Flat" space. The configuration of our space is
  strait, and no matter the speed of the objects, they always are
  traveling in a straight line.
- In diagram "B" is shown, how two objects with different speeds will
  travel in curved shape space. - No matter the speed of the objects,
  they must follow strictly the shape of the space because this shape
  and curvatures are their "strait line" of travel. It is obvious that the
  two objects will always travel in parallel lines regardless of their
  speed.
- Einstein is claiming that Gravity does not exist, but the shape of the
  space is producing this effect. The best test for the correctness of his
  claim is to observe two objects with different speeds when they are
  passing near a massive object. If the massive object is curving the
  space, and gravity do not exist, the two traveling objects must travel
  in parallel lines to each other and strictly follow the curvature of

space, but if the space near the massive body is not curved, but is strait and gravity really exist, then we will observe that the gravity will bend the path of the slow traveling object much more than the path of the fast traveling object! - And this is exactly what we are observing in our physical world! - (See diagram "C").

That's why we have the three "Escape velocities" for the Solar System because the traveling objects tend to travel in a straight line, regardless of the presence of massive objects.

I believe that this example is undeniable proof of the existence of Gravity as a real existing force. It is easy to be understood by everybody, and this understanding will put out the speculations for the "Curvature" of space once at per all!

The inherited incorrect-expanding and open space model of The Big Bang theory and Einstein's weird assumption for space curvature create a real physical mess and are stripping us of the chance to understand the real nature of Gravity and what the attractive mechanism and forces they are. Those fundamentally incorrect configurations are leading to the creation of endless weird mathematical models and theories, which do not represent reality, do not make sense, and are not explaining anything – as gravitons, gluons, and the "discovery" of Higgs boson!

In reality, there is nothing mysterious or complicated to understand how the attractive forces and gravity works. It is not very difficult to comprehend if we keep our feet on solid ground and using the law of Physics, proven facts, and observations to build a correct model and correct configuration of the fundamental building blocks of our Universe. There is only one attractive mechanism in the Universe - it is based on the Law of Physics, which states that energy and matter are moving from a state of higher energy level towards a lower energy level. - (that is it! - no more, no less!)

Now we have to find if our model can explain all the weird properties of Gravity. Our explanation has to be simple and to be within the boundary of the Law of Physics. If we are able to do this, this will be undeniable proof of the correctness of our structural model of the Universe.

Until here this was part of my incomplete explanation of Gravity my friends because I was afraid to revile the source and the property of Gravity. Don't be disappointed to learn that the biggest mystery in Physics is a very simple mechanism:

Gravity is a fundamental force that attracting equals all forms of matter regardless of its polarity. This uniform attractive property of Gravity is a vital clue for its origin and is an actual direction where we have to start looking to find it. This is logical, and that's why I would like to start the explanation by start looking deep into the structure of matter - into the first building blocks of matter :

We know that the solid-state of particles are standing waves around the center of the energy deficit. This configuration is providing a condition, where the binding forces are well balanced and are confined in the perimeter of the particle. For simplicity, and to be able to explain the concept, we will ignore the other properties of the particle and will concentrate on those two opposing each other forces - the standing wave, and the energy deficit in the center. More or less, this configuration is acting as a closed physical system, where the internal binding force is locked and is not leaking out. For easy understanding, we will consider this closed physical system as one cocoon or sphere with an outer surface, or "membrane". The "membrane" is the surface of the elementary particle, where it is in contact with the surrounding space. Exactly on this surface, gravity is produced and exists! - Despite that, the binding force is a confined force, <u>the centrally located energy deficit is engulfing and exceeds the boundary of the standing wave</u>. As a result of this, the outer layer of the Particle has a slightly negative value. - Exactly this is the reason for Gravity! - Space is a physical medium, and the touching layer of space inevitably will have (lower) - non-zero energy level! - <u>The first, the touching layer of space inevitably will have a lower energy level than space behind because it is in contact with a system with an energy deficit! -</u> **This is the mechanism of Gravity!** - That the binding force of the elementary particles is producing (energy deficit), or non-zero energy level on the touching space surface. The binding force of particles is not leaking out, because, when the particles are formed, the <u>Gravity a space energy imbalance is formed on the surface of each particle and stays there</u>. This space energy imbalance is locked between the surfaces of particles and the touching layer of space as <u>confined energy in the structure of space</u>! It is similar to the surface tension of molecules or water surface tension - it is there, it exists and it stays there! The Universe is a closed physical system, and the inserted energy in the form of Gravity has nowhere else to go. In practice there are no particle energy losses because this surface energy imbalance (deficit) is

constant it stays there! <u>It becomes an energy imbalance of space</u> and is locked
around each particle. That's why Gravity is proportional to the mass of the
object. That's why Gravity is acting in a normal way as emitted force, but it is
not for the expense of the internal energy storage of the elementary particles
The energy imbalance (or energy deficit) is caused by mass, but is locked in
space! The energy saturation or (space or energy "pressure") is pushing the
material objects together, but there is no energy gain or loss because Gravity
is confined energy between matter and space. In the case of gravity, space is
seeking to restore equilibrium by pushing the objects together.

For better understanding I will provide the example with a permanent magnet
- When the magnet is attached to the fridge its magnetic field gets locked
between the steel and the magnet – the magnetic force becomes (confined),
and the magnet is not losing its energy! If the magnet losing its energy, it will
stay there only a few days before falls off, but the magnets stay attached for
many years and are not losing their magnetic attraction. We have to
understand the difference between a confined configuration of energy and
loosed configuration. - This is the reason why the particles and atoms are
stable for billions of years because their energy is in a confined configuration.
<u>Gravity exists on the same principle, it is confined energy</u>, just on a bigger
scale of the Universe, but the Universe is a closed physical system and the
energy stays there! - The energy can be loosed in one place but will appear in
another. We have considered the Gravity of particles. Let see now what is the
situation with the atomic nucleus and composite particles like protons and
neutrons and if there are conditions for Gravity to be produced and exist?

The strong nuclear force also is (confined space energy deficit) in the center of
protons, neutrons, and the atomic nucleus, which is acting as an attractive
force. - It is confined energy in the range of the composite particles and the
nucleus. The strong nuclear force is a strictly confined force in the range of the
particle or nucleus, but on the outer surface - ( the contacting surface with
space) has a negative - non-zero energy value.

This negligible surface energy deficit created by the strong nuclear force is
also a source of Gravity. It is negligible, but with a gathering of mass, the
gravity is increasing its strength and is able to hold together all celestial
bodies. In practice, gravity is acting as a "radiated" force in space similar to
electromagnetism, because <u>is not confined in the matter, but is stores as a
property of space</u>. <u>The neutral nature of the strong nuclear force makes the</u>

Gravity also to be a neutral -(non-polarized) attractive force.

We have to continue our consideration of the biggest structure of matter to see are they not contributing also to the force of Gravity.

The next level is the atom by self. We know that the atoms are held together by a similar configuration of confined electromagnetic attraction between the protons and electrons. In this configuration we know, that the well-balanced electromagnetic attraction providing conditions for the atoms to be considered also as a closed physical system, because we know, that atoms can exist for billions of years without losing their energy and internal bond. The polarized forces of protons and electrons are in balance, and as a unit, the atoms are electrically neutral. Armed with this knowledge, will be easy for us to realize, that each atom also will create the same non-zero energy value on the first touching layer of space - the same as the elementary particles and the atomic nucleus. This energy imbalance on the outer surface of atoms will be 137 times weaker than the energy imbalance of the nucleus because the electromagnetic force is 137 times weaker than the strong nuclear force, but there is a lode atom in the Universe and their combined force is significant!

To be able fully to understand the "puzzle" of Gravity, we have to continue our consideration of the bigger structures of matter which is saturating the interstellar medium. They are the molecules, ions, plasma, charged particles, and dust. From the consideration above, we know that every center of attraction will produce gravity. In the case of the mentioned constituencies of the interstellar medium, their attractive force is electromagnetism and static electricity. We have mentioned above, that electromagnetism is 137 times weaker than the strong nuclear force, but this is not something that we can ignore when this is on the colossal scale of the Universe! The realization that electromagnetism can produce gravity is shading light on the currently unsolved puzzle of the rotational unity of the galaxy, the strong gravitational presence in the centers of extreme electromagnetic phenomena's as Quasars, Pulsars, and the galactic centers, where the incredible strong electromagnetic forces are mimicking the presence of supermassive black holes. I am not suggesting that at the center of galaxies is not massive material objects, I just would like to explain that we still cannot differentiate between the gravity of the matter and the created gravity by the extreme electromagnetic forces. The same is the situation with the assumed mass and distance of Quasars. The fact that Quasars can be ejected by the galaxies telling us that they are not very massive objects and their strong redshift is a result of the extreme electromagnetic force present there.

In the case of the unresolved "puzzle" of quick star formation, the knowledge that electromagnetism is producing gravitational attraction is enough to shed light on this currently unresolved problem in astrophysics.

What about antimatter?

We know, that everything else a part of the opposite polarity of its particles is the same as normal matter. That means that antimatter creates gravity in the same way as matter and radiates the same non-polarized "positive" gravity as our ordinary matter!

As every phenomenon without mass, gravity must travel not slower than the speed of light. However, we know that photons effectively have mass and this fact putting big question marc on the upper limit of travel. Allegedly, the neutrinos are slightly faster than photons. By using this knowledge, we can expect that gravity also will propagate faster than light and faster than neutrinos, but how fast, this will tell the future. I believe that the speed of Gravity will be the ultimate limit of travel in the Universe!

It's turn out, that Gravity is acting as a real irradiative force, but is not leaked energy out of matter and particles, but is an existing energy imbalance - and it is property of space! - It is a confined space energy deficit between space and matter. It is part of space energy and is created simultaneously with the creation of matter.

There has been a big controversy at the beginning of twenty century with the Biefeld-Brown Effect discovered in 1923. The authors of this effect are claiming that they have produced gravity and antigravity. The well-lubricated machine of academic science manages to suppress this news quickly and no further experiments and publications have been allowed. Despite the existence of many patents for antigravity mechanisms, the academic machine of suppression has declared the Biefeld-Brown Effect to be just an ionic wind, which is mimicking Gravity.

The problem for the establishment comes from a Russian-born scientist - Dr. Eugene Podkletnov who was working in the USA and England and has done experiments that clearly show that electromagnetism is producing Gravity. In 1995 when he tries to publish the results of his work, his publication has been blocked, his working contract was canceled. He has been forced to keep quiet. Despite the brutal repressions Dr. Podkletnov found a job in a university in

Finland and continues his experiments privately. He has done the experiment in a vacuum chamber with the same result. This is proving that he has producing Gravity and the hypothetical "ionic wind" is just a smokescreen for the establishment to hide this vital knowledge from the public. Dr. Podkletnov has produced also a directional gravitational impulse, capable to punch thru thick steel plates and concrete walls at a distance of up to 5 km. He has measured the speed of the gravitational impulse, and he claims that Gravity is traveling 64 times faster than the speed of light. Despite his practical achievements, Dr. Podkletnov hasn't been able to give a theoretical explanation of Gravity and the Biefeld-Brown Effect.

Thanks to the brave Dr. Podkletnov, now we have an experimental proof for the correctness of my theory, which is the theoretical explanation of the Biefeld-Brown Effect and the experiments of Dr. Podkletnov.

All these experiments are confirming my findings, that there is only one fundamental force in the Universe, and that the energy deficit is the universal principle of attraction.

If we would like to be able to produce gravity and antigravity, we have to find a way to produce energy deficit and energy excess into the required point of space - for example - below the spacecraft or on the side of it. - This will be the most efficient propulsion system, which will work in the same way in the air, underwater, and in space too,

Not many understand that with gravity propulsion the pilot of gravity propelled craft will not fill the effect of rapid acceleration or stopping because the gravity is acting also on each atom and molecule of the body of the pilot in the same way as to the craft. If the claim that Gravity travels 64 times faster than light is correct, this will open the door for humanity to travel to the nearest stars faster than our current ability to reach Mars.

Currently, we are restricted from the study of Gravity from the military secrecy and mainly, because there is a well-entrenched and scientific dogma of the "Theory of Relativity" and the "Standard Model" which is not allowed to be questioned.

To be able to master the property of Gravity, first, we have to put our knowledge of correct fundamental principles and start using logic. We need to use scientific facilities to study the correct and the real property and interactions of matter and its related forces.

## MATTER - ENERGY EXCHANGE AND THE RECYCLING MECHANISM OF THE UNIVERSE:

The ancient philosophers have achieved great understandings of The Universe and the structure of matter. Their scientific data has been very limited, but by using intelligence and logic they have been able to understand the Heliocentric Solar System and by using the size of the shadows of the solar and moon eclipses to calculate the distances and sizes of the Sun, Moon, and Earth. By using observations of chemical reactions, they were able to construct a very modern atomic model of matter. They have been intelligent enough to understand also that principles of ethics and democracy should be the fundamental principles of an intelligent society. All those amazing achievements the ancient philosophers succeed to make because they were using logical analysis of the available facts!

Currently, we possess an enormous amount of scientific facts and data, but in general, we are far behind our ancestors in the understanding of the World. - Why?

The answer is simple; we have departed from common sense and logic in every aspect of our life, especially in science! Mainstream science considers only ordinary matter and gravity to explain the Universe, but the problem is there, where 99% of the Universe is in the form of plasma, and plasma is ruled by electromagnetism. Electromagnetism is many billion times stronger than gravity and shouldn't be ignored! By considering only 1% of the existing matter and ignoring the strongest and dominant electromagnetic force, the present science will never be able even to get close to an understanding of how the Universe works! - This is not a very pleasant situation, and we have to address it with care, and logical consideration of the available facts!

To be able to construct a realistic model of the Universe, we have to include all known existing elements, without preferences and ignorance!

In interstellar space, there is spread a huge amount of defused matter in the form of clouds of dust, plasma, gases, ionized particles, molecules, and neutral atomic matter. (See the image on page 6)

Despite the lack of density, this diffused matter is representing most of the matter of the Universe, because the distances between the stars and galaxies are enormous. The other phenomenon, which is not considered by the current model, is the electromagnetic property and interactions of the celestial structures. Everybody knows that Earth has a magnetic field that keeps the compass needle facing north-south and protects us from harmful solar and cosmic radiation. Every star and most of the planets also have magnetic fields. Each galaxy and each structure of the Universe also contains magnetic fields. The Universe is a dynamic place, where everything is moving, shifting, and rotating. The Earth's magnetic field is rotating in the Solar magnetic field. The Solar magnetic field is rotating in the Galactic magnetic field. The Galaxy is rotating in the magnetic field of the Universe.

We know that when magnets or magnetic fields are rotating in another magnetic field this rotation produces current – (the dynamo effect)! The intergalactic currents are producing enormous magnetic fields and a cosmic plasma web! Electromagnetism producing a bonding effect to the galaxies - similar to gravity and also has unlimited range! Electromagnetism is energy, which is acting also as mass! - The current science is not taking this into account. The electromagnetic forces are many billion times stronger than gravity - ($10^{38}$). Some electromagnetic processes in the Universe possess enormous strength! They are able to produce up to 40 million times more energetic beams than our best particle accelerator! The magnetic field of a quasar reaches 200 million times the strength of the earth's magnetic field! After 35 years of traveling, when Voyager spacecraft left the Heliosphere, the scientists with disbelieve discover that there exists a "Firewall" with the electrical potential of billions of volts! This "Firewall" is the meeting point of the Solar and Galactic magnetic field. The diameter of the Heliosphere is enormous - 36 billion kilometers and the electrical current coming from the Galaxy into our Sun are enormous!

The dynamic electromagnetic interaction of the moving and rotating structures of the Universe is the main mechanism behind the distribution of energy and matter between galaxies and stars. Electromagnetism is eroding the dying stars and they are releasing enormous amounts of charged particles back into space, where the electromagnetic currents ferry and distribute plasma and diffused matter to the regions where the new stars are forming. Electromagnetism poses attractive and repulsive power and these properties are keeping the balance of the mass concentration in the Universe. Increasing the mass concentration is increasing also the energy level of this region. The law of physics dictates that energy is moving from the higher level toward the lower level. The streams of plasma are stretching and connecting the Universes structures and transferring the matter particles on long distances.

The charged particles are dragging the neutral particles and these combined streams are feeding the galaxies with the necessary matter for the new stars formations. Those streams of diffused matter have been invisible in the past, but this is not the case for our new sophisticated instruments. We are observing that the Galaxies are situated on the strings of the plasma filaments and their orientation and rotation are dependent on the direction of the electromagnetic current of the plasma filaments. Further, electromagnetism is providing and regulating the rotational speed of the celestial bodies. In this way, the dynamic of the system is maintained.

The work of the Universe is based on a very simple, but very reliable principle – balanced dynamic and continuous energy-matter exchange. To understand how the system works, we have to understand how the structure of the Universe is designed. The fundamental elements of Universe structures are Space, Matter, Time, and the Laws of physics. We mentioned above, that for every quantity of matter there is an allocated specific volume of space! That means, that the Universe is acting as a closed physical system, (regardless of it is endless or finite), where space has constant volume, and there is an inserted specific (proportional) amount of energy (matter). - This inserted energy has nowhere else to go - soonest is released from one structure, it is available to another! - Our observations show that hydrogen is the most abundant element in the Universe. Hydrogen is the most energetic element of all elements! When the hydrogen starts fusing into heavier elements the nuclear fusion releases enormous amounts of energy into the space of the Universe! This released energy is boosting and is increasing the rate of the reverse process - which is the electromagnetic fission of heavier elements. This reverse process is restoring the balance ratio of hydrogen/heavier elements; the amount of diffused matter and the saturation of interstellar space with thermal and electromagnetic energy. - (Recently, the "Sapphire Project" has confirmed the synthesis of elements in strong and hot plasma interactions in conditions close to the Sun corona).

The Universe is possessing enormous systems of dynamic plasma distribution, which provides material for continuous matter-energy exchange between all celestial structures. The magnetic field and the gravitational motions of celestial bodies are producing enormous electromagnetic currents, which is forming powerful electromagnetic "vortexes" (accretion discs) – which are mistaken to be Black Holes. There the released energy is breaking the elements back into hydrogen! The unexplained abundance of neutrinos coming from all directions is the actual proof of the continuous recycling processes of the Universe. -

And this exactly is the self-balancing mechanism of the Universe, - the fine-tuned ratio between (**fusion and fission**) - the balance ratio between the released energy of **hydrogen fusion** and the **quantity of heavier elements**! - As soon as the amount of heavier elements gets increased by fusion, the released energy increasing the amount of interstellar electromagnetic energy with geometric proportion and boosts the reverse process of the electromagnetic **nuclear fission** which is breaking the heaviest elements back into hydrogen and is stabilizing the precisely chosen balance-ratio of the three components: - heaviest elements to hydrogen, and the saturation of interstellar space with thermal and electromagnetic energy. - And this exactly is the picture, which we are observing - One vast, incredible and beautiful Universe, balanced with absolute precision, stability, and simplicity, where the fine balance of forces and kinetic energy provides continuity for the eternal dynamic existence of all the structures of the Universe!

Someone could ask me why are no mathematical formulas in my Theory? The answer is simple: the mathematical formulas have no meaning! Scientists have tried for millenniums to explain the World with mathematics, but they didn't succeed. The Universe is built with logic, harmony, and purpose. - Just ask yourself: what we see, what we hear, what we fill, and what we know about the World around us? We see the beauty of nature in colors. We hear the sound and the harmony of music; but why the colors are seven? Why the notes are also seven? What makes the colors and notes come in harmony with Nature? We are part of the Universe and part of Nature! We are filling the World exactly how it is! We know that there are six world directions; we know also that Time exists! - Space and Time are the Seven Dimensions of the World. The sound and light are vibrations, and only when they are in phase with the Seven Space Dimensions, only then they become in harmony with Nature. That's why we are enjoying the harmony of music and the beauty of Nature! With our eyes, our fillings, and our senses we can understand the World much, much better than any mathematical formula! Our ancestors tried to give us knowledge and explanation - they tell us that the matter of the Universe has been created in Six days and the seventh day is a Time for rest! Now we can understand what they were telling us, but do we really want to know the truth?

Mathematics is not providing explanations! If you want to understand the World, put the mathematic aside and open your eyes and fillings! - The World is out there for you to fill, to explore, and to enjoy its beauty!

It turns out that the Universe is enormous! Much, much bigger than academia suggested! It has no beginning, or at least not the beginning which the current "scientific" theories suggest!

There will be no cold dark end of The Universe, which the inflation theory proposing because the Universe is constant and is here to stay! The Universe is constructed with vision, logic, precision, and self-balancing mechanisms where the matter-energy exchange provides the rejuvenating and dynamic eternal existence of the celestial structures!

I believe that this result will give the answer and satisfaction to all intelligent and honest people, which will be happy to know the truth about what kind of World we are really living in and what the future of the Universe really is!

## UNDERSTANDING THE CELESTIAL BODIES AND STRUCTURES OF THE UNIVERSE:

### Formation of Stars and Galaxies

The Standard Model of Cosmology is giving us a very unrealistic mechanism for the formation of celestial bodies. This model is driven by the materialistic philosophy adapted by academia, where all the processes in the Universe are the result of mechanical interactions driven by Gravity. Such a primitive approach cannot produce a realistic model of the formation of the celestial structures. The model is proposing that under the influence of gravity, the gas molecules in space are attracting each other and are forming gas clouds and nebulas and gravity make the cloud denser and denser and finally is collapsing under the influence of gravity. From physics, we know that the natural tendency of gases in a vacuum is to disperse. This has been tested thousands and thousands of times and results are always the same - The gas molecules are dispersing in a vacuum toward infinity and gravity has no effect on them. Even if the slightest condensation of gas molecules occurs in some area, the passing radiation is heating the gas molecules and the heat is acting as a powerful dispersing force.

If we start considering the electrical charges of gas molecules, then we can see that the passing electromagnetic radiation is knocking off electrons and is ionizing the molecules and particles of the gas clouds. We know that the electromagnetic attraction between charged particles is a billion times stronger than the gravitational attraction between these particles.

To understand what mechanism is forming the plasma and dust clouds in space, we have to observe them and take into account all forces presented in this process regardless of the dogma inserted from the political establishment. Knowing this, we have to face reality and point out radio telescopes toward the molecular clouds and nebulas. - We have done this, and the results of our observations are undeniable! - In each molecular cloud and nebula is presented a substantial electromagnetic field, which is the driving mechanism of gas condensation and formation of stars and planets! Our observations is confirming the dominant influence of electromagnetism and we seeing that the galaxies are situated on the strings of the cosmic plasma filaments and their rotation is dependent on the direction of the electrical current of the particular filament. Also, electromagnetism is responsible and is regulating the rotational speed of the celestial bodies. It is not a good idea to ignore the obvious and to put our heads in the sand and continue to maintain an impossible materialistic mechanism of star formation.


## Sun and Stars

Out Sun is the most important element for the life on Earth. We are enjoying the sunshine every day, but we not asking ourselves do we really understand what the Sun is and how the Sun manages to produce constant and steady energy for billions of years?

I will start with a brief explanation of the accepted model of the Sun:

To be in line with the theory of the Big Bang, the leading scientists declare that our Sun must be formed of "only available" at the time material - 75% Hydrogen and 25% helium. They declare that the Sun is a giant ball of gas. Under the enormous pressure and temperature in the Sun's center, the temperature reaching 10 to 15 million degrees and Nuclear Fusion has started there. The produced abundance of photons is producing "radiation pressure" which is keeping the balance of the system from collapsing. Some self-balancing mechanism is preventing the Sun to explode as an enormous hydrogen bomb. They suggest that Hydrogen and Helium content will be able to produce energy for about 10 - 15 billion years, and when the fuel is exhausted, the Sun will become a red giant or will explode as a supernova.

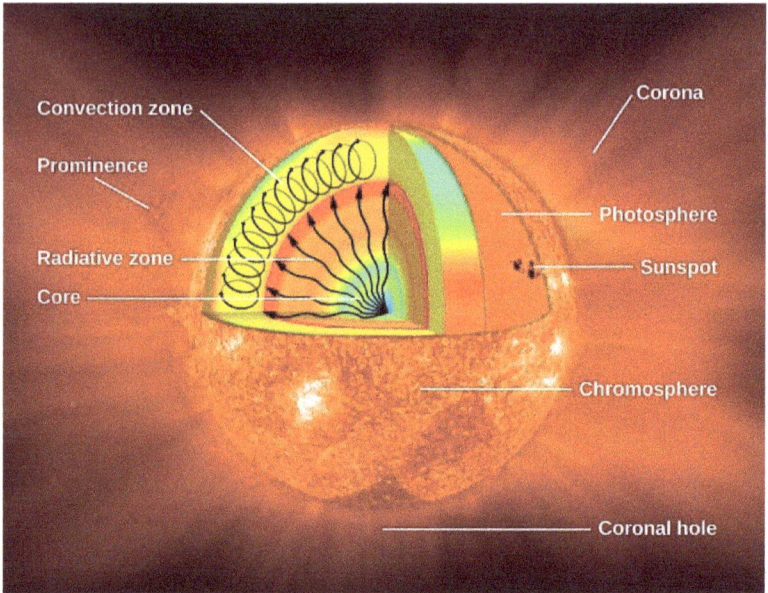

This is our current model of the Sun

To be able to have confidence in our judgment, we have to understand well what the established criteria are for a scientific theory is to be correct. - Recently, the Higgs Boson to be accepted was necessary to produce results with certainty to five decimal points, which suppose to be an accuracy of (three and half million against one) certainty.

I will give one example that even the most insignificant anomaly of scientific theory will be crucial and will prove that this theory is not correct. - The Geocentric model of the past. Everything was obvious and logical in this model because everybody can see with his own eyes that the Sun is circling the Earth. There was only one insignificant anomaly in this model - there were five stars, which was changing chaotically their path and position in the sky - (the visible planets). This "insignificant" anomaly was ignored. -  According to the modern standard of five sigma's, the Geo-centric model still is well within the boundary of the standard of acceptance, because the ratio of five moving stars to the 200 billion steady stars is well above the required five sigma standard... but is an anomaly!

Thanks to our sophisticated telescopes, now we have the confidence to understand that even such an insignificant anomaly in scientific theory is not acceptable and that any anomaly regardless of how insignificant is, it is proof that the theory cannot be correct!

Armed with this understanding we can start the consideration of the available data and accumulated facts to see if the officially accepted and promoted

Sun's Model is correct and if there are any anomalies that we have to accept or reject. The theory and model of the Sun were produced a long time ago when astronomy was in its infant stage of development and the astronomers had not enough information for the structures, dynamics, and composition of the celestial bodies. During the past 60 - 70 years, our telescopes and instruments have become very sophisticated and we manage to collect an enormous amount of data for the structure of the Universe. So far, mainstream science is very comfortable with the established theories and is not willing even to start questioning them. The problem is there, where many new discoveries cannot be fitted in these theories and are creating too many unacceptable for scientific theory significant anomalies. I believe that it is time to re-consider the available facts and to readjust our understanding of the Sun, stars, and the Universe.

The first sign that something is not correct with the current model of the Sun comes from the attempts to re-create the conditions of the Sun's interior and to produce Nuclear Fusion. The famous "Tokamak" has managed to provide the required millions of degrees temperature and pressure, but Nuclear Fusion hasn't been achieved. Already six decades all industrial countries are following this way and are trying to produce Nuclear Fusion, but no matter how hard they try, the result remains elusive and are always twenty years away. Why? - The obvious answer is that the approach is incorrect because our understanding of the Sun is not correct.

Mainstream science is giving us a Solar Model with an impossible property! - It is impossible because the claims of the model not only don't match the available facts and observations but the proposed temperatures of the model are in the reverse sequence of what we are observing and measuring on the Sun. To cover up the nonsense of Big Bang theory, which dictates, that the stars must be formed from only available "primordial" staff – hydrogen and helium, the mainstream science gives us the model, where the Sun is supposed to be just a ball of hot gases, but the facts show that this explanation is absolutely incorrect. Here are the facts which are in total conflict with what they are telling us:

The model claims that the Sun is a hot ball of gases - mainly hydrogen. Nuclear fusion in the Sun's core is producing heat. Some invented "radiation pressure" of "mass-less" photons preventing the Sun to collapse. They also invented some "mysterious" mechanism that prevents the Sun to explode as one ordinary enormous hydrogen atomic bomb.

One of the questions is - how the "mass-less" photons will be able to produce "radiation pressure" which will be able to outstand and contra-balance the billions of tones of Sun mass and pressure? They have invented this "radiation pressure" to cover up the nonsense of the model. Everybody can check it up and make a simple test - just put a stronger light in a balloon, take the air out and seal the balloon. Then switch on the light. If the balloon starts inflating, that means that the claim for the "radiation pressure" is correct, but if the balloon is not inflating, that means that the claim for "radiation pressure" is no more than an ordinary deception.

The current model of the Sun dynamics is based mainly on gravity and pressure. In this gravitational model, many properties and interactions are labeled as "Puzzles." The problem with the proposed model is there were our observations show a completely different picture of what they proposing. - The heat distribution of the Sun layers is in a reverse sequence of the proposed model. One of the "Puzzles" (for them) is that the Sun has a surface and atmosphere! It is a well-known fact that gases cannot have a surface and atmosphere! The real temperature sequence is in the opposite order - the visible internal sun temperature visible through the sunspots vortexes in the depth of thousands of kilometers is 4,000K, followed by the Sun's photosphere 5,800K, and above it is the Sun's Corona with temperatures up to 2,000,000K.

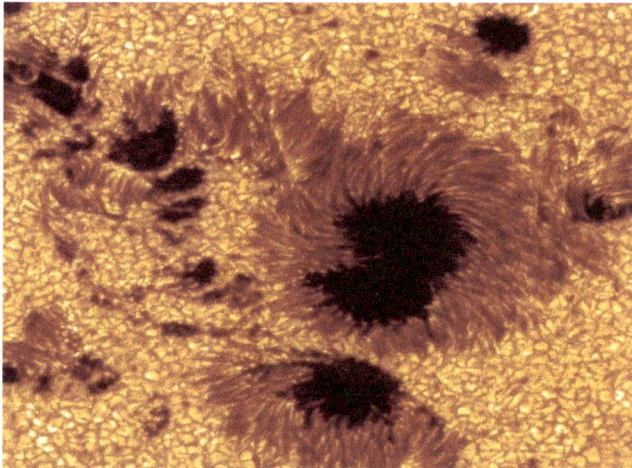

Photo of Sun surface and its cooler interior.
But is the "gas giant" doesn't suppose to have a surface and atmosphere?

The sequence of convectional heat exchange mechanisms of the proposed model is exactly in the opposite order – (the Sun interior must be hotter). The official Sun's Model cannot explain such opposite and bizarre heat sequences.

Therefore, the creators of the gravitational, nuclear fusion Model of the Sun declare again that this anomaly is a "Big Puzzle"!

The known fact also is that If the star's interior for some reason reaches the necessary conditions for nuclear fusion (how the model is claiming to be), nothing will stop the star from exploding as one ordinary hydrogen bomb, but the next "Puzzle" (for them) is that the Sun is not exploding!

To be able to understand what really is going on, we have to start with consideration of the physical state and functions of the solar mass. Mainstream science claiming that the Sun is just a hot ball of gases, mainly hydrogen, but the numbers don't add, because the Sun's specific gravity weight is 1.4 metric tons per cubic meter! – To compare this number we have to know that the specific gravity weight of Hydrogen is 0.007L only! The liquefied hydrogen is 0,07L. Even the metallic liquid hydrogen is 0.7L. All those three figures are far below the specific gravity weight of the Sun, which is 1.4L! - The difference between those two figures is on a magnitude of 20 to one, which is equal to an unacceptable error of 2,000%! That's why I have started with the statement for the absurdity of the official Solar Model because even the specific gravity of metallic liquid hydrogen is 0.7, which is 700g/L. This also is far below the Sun's specific weight of 1.4kg/L.

From physics, we know that no matter how much we are compressing gases, they always remain lighter than their liquefied state! From physics, we also know that we cannot compress liquids - their volume remains constant, no matter the pressure we apply. From those known and undeniable facts become more than obvious that the Sun cannot be just a ball of hot gases and even cannot be just liquid or solid hydrogen! - The Sun is much, much heavier than that! Unfortunately, they are hiding and avoid mentioning these facts. Under enormous pressure, most of the interior of the Sun inevitably will form liquid metallic hydrogen with a solid metallic core, because it is obvious that the Sun's interior is not a million degrees hot, but is only 4,000K! Opposite to the other gases, hydrogen needs a temperature over 1,000C to break the molecular bond and to transfer into liquid metallic form. This strange physical property of hydrogen also looks like to be carefully "chosen" to support the energy exchange of the stars – a similar case we observing with the carefully chosen physical property of water - it expands when freezing! Scientists suggest that metallic liquid hydrogen even at this high temperature acting as super-fluid and superconductor! - This explaining why the Sun's interior is cooler than the surface, because of its heat superconductivity. The enormous Sun's magnetic field also supporting this claim and is a clear indication of the fluid activity of the liquid metallic core. The ability of the Sun's structure to absorb, transform, and emit energy is amazing! Such a structure will be presented in every star, and this structure obviously cannot collapse, how the "model" is claiming. Liquid and solid materials are not collapsing! The assumptions for collapsing of stars are just baseless fantasy, which they created, to be in line with the age of the Universe, proposed by the Big Bang theory.

The understanding that a star's interior structure is formed not of gas, but is mainly liquid with a solid metallic center is ruling out the possibility of gravitational collapse of the star. This understanding is ruling out the assumed result of such gravitational collapse of stars - the weird theories for the existence of Neutron stars and "gravitational star collapse" - producing Supernovas and Black Holes! The reason for the exploding stars is absolutely different than the officially promoted ideas, and we will consider it in another chapter.

The conditions for thermonuclear fusion, according to the scientists, need temperatures above 10 million degrees but the Sun's internal temperature is a miserable 4,000 to 5,000 degrees! Question is - how is it possible for the Sun's surface to be 5,800 degrees only if is sandwiched between two layers millions of degrees hot? It is obvious that this is an impossible claim, and we can see and know that Sun's interior is 4,000k.

The neutrinos coming from the Sun are not enough to support and explain nuclear fusion. - This is another indication that the current Sun's model is incorrect! But the guys are clever! They somehow "discover" that the neutrinos coming from the Sun are "enough" to explain nuclear fusion because on the way from the Sun to Earth the neutrinos are "changing flavor"! - Brilliant! Congratulations! They even discover that neutrinos have mass! Fantastic! - Everybody knows that there is no such thing as mass-less particles because the particle has to be made from something! - And for this "genius" discovery "the guys" give themselves a Nobel Prize! ("They" must be very, very intelligent). But... there is a problem...! "Small problem"... - The neutrino detector hasn't been situated between the Sun and Earth, but underground! - How do they detect, measure, and decide what kind of neutrino is coming from the Sun and where they are changing the flavor? Through every square centimeter surface, every second is passing billions of neutrinos coming from space! In the huge neutrino detector, they are detecting only about ten neutrinos a day! What the statistical credibility of such a result is? – Obviously, Zero!

Electromagnetism is a force that is $10^{38}$ stronger than gravity, but the mainstream scientists are stuck stubbornly to a gravitational model and are refusing even to consider the electromagnetic processes of the Sun! - Why? For the strength of the Sun's electromagnetic field, you can judge from this fact:

The Sun gravity is enormous, and despite this, the electromagnetic corona mass ejection ejecting and accelerating billions of tons of hot plasma in a matter of seconds up to ¼ of the speed of light into interstellar space! – Is this fact deserve to be ignored just because not fits in their theory?

The Sun activity is on a regular basis of the 22-year circle, where every 11 years the Sun is changing its polarity! - This cannot be explained with nuclear fusion, but with electromagnetic interaction based on a regular rotational cycle of the Sun into the superior Galactic electromagnetic field.

This is only a "small" example of the real pile of obvious miss-assumptions, and unexplainable phenomena that can be explained easily with electromagnetism, but not with gravity, pressure, and baseless fantasy.

When we observing the sky, the position and the orientation of the Galaxies is looking chaotic, but with the aim of radio telescopes, the scientists were able to see the interstellar plasma filaments and the biggest surprise was that most of the galaxies are lined up on the "ropes" of the plasma filament and are always perpendicular to the direction of the plasma filament! This fact is enough to be realized that intergalactic electromagnetism is the crucial element in the formation of the celestial bodies and their energy production.

The mainstream scientists are assuming that Sun is just a ball of hot plasma, but plasma is electrically charged particles and molecules which obey much more electromagnetism than gravity but "they" just ignoring this fact and are giving us the absurd scenario of gravity-driven Solar Model. I can say only - that ethics and honesty should be a priority of science!

When we are looking at the real footage of the loops of plasma above the sunspots, we can see by ourselves that when they burst, the plasma streams are accelerating away from the Sun with continuous increasing speed! - This continuous acceleration cannot be explained by gravity because it is an obvious sign of the continuous electromagnetic acceleration of billion tones of hot plasma in a matter of seconds in the exact opposite direction of the Sun's gravity.

It is well documented the presence of substantial Sun and Galactic electromagnetic fields. Those two well-known electromagnetic fields – (Sun's and Galactic) have to interact at some point where this interaction inevitably will produce heat! - This is exactly the phenomenon that we are observing in the Sun's Corona! - Interaction and canceling of magnetic fields! Those magnetic fields are enormous and the steady and continuous energy of the

Sun is a result of their interactions! - I am sorry my friend, but we have to accept the facts!

Here is an example of the absurd of this situation:

If I try to explain the function of the microwave oven and the arc welder with gravity and pressure, you will label me as stupid and crazy! Why? - Because it is obvious and well known that the heat of the microwave oven and the arc welder is produced by electromagnetism and gravity has nothing to do with it A very similar situation we have with the Sun, but in this case, everything is in reverse order of observational data, facts, knowledge, and logic! The people who are explaining the plasma interaction of the Sun with electromagnetism are labeled as uneducated. Why? Who is the uneducated, the one who explains the interactions with facts and the law of physics, or the one, who is ignoring all the facts and declares them a "Puzzle"?

The attempt to slide under the carpet all those facts and proclaim them as a "Puzzle" is a very disturbing way to do science!

In the beginning, we have considered what are the criteria for some scientific theories to be correct? We have reached the conclusion that fundamental scientific theory must be clinically correct and not have room for any anomaly or "Puzzles". Unfortunately, when we considering the available facts and observations become obvious that the officially accepted model of the Sun is full of discrepancy and weird claims which are exactly the opposite of the facts. This model is full of baseless assumptions which not reflect reality. Such kind of false knowledge is not of value but is an obstacle to progress. Will be good if we are able to collect courage and to say: enough, we need to know the truth!

### Supernova - what is it?

Supernova is the last and the final stage of some stars. Some stars ending their life with the most powerful explosion in the Universe, which we calling a Supernova. There are five types of Supernova and their classification is based on their luminosity, duration, and absorption lines of their emission. Type 1A Supernovas have always the same brightness and are chosen as "standard candles" to measure the vast distances of the Universe. We will not go deep into the technical details of the property of Supernovas. We will consider just the main concept of the officially provided model of the Supernova and the available observational and scientific data.

I believe that for the educated and intelligent people will be of great interest to know what the reasons are for such a sudden and violent end of some of the stars. The advance in astronomy is giving us many new facts, which is not fully supporting the officially accepted model for the composition, the physical processes, and the life and death of the stars. The reason for the mismatch of the observational data and the theory is coming from the desire of the mainstream science to adjust the process of stars formation to the theory of the Big Bang where the stars can be formed from only available material produced from the Big Bang, which supposes to be 75% Hydrogen and 25% helium. There should be a few percent of other elements, but their amount is negligible. Mainstream science is assuming that when the star is formed, under the enormous pressure and temperature in the star core starts Nuclear Fusion of the hydrogen content of the star is turned into helium. Nuclear Fusion supposes to be the energy source of the stars. The produced heat and the radiation pressure of photons are keeping the balance of internal star pressure and are preventing the star from gravitational collapse. This process lasting billions of years, but eventually, the hydrogen one day will finish. When the hydrogen is finished, then the star gets a bit hotter and large as (red giant) and starts fusing Helium into lithium and into other elements as carbon, oxygen, and silica. This process also lasting relatively longer, but one day when the lighter elements also finish starts production of iron. The problem is there, where the nuclear fusion of iron is not producing energy but is absorbing energy. This energy absorbing reaction suddenly disrupting the balance of the internal and the external gravitational pressure and the star suddenly collapses. The hurling in collapsing matter creates an enormous shock wave, enormous pressure, and billions of degrees in a fraction of a second. - This is the official explanation of how and why some stars are exploding with enormous power as one giant atomic bomb. This exploding star we are calling - Supernova. Under the enormous pressure and temperature of the explosion, the matter particles of the central region of the exploding star are compressed by billions of atmospheric pressure, and as a result of this; the electrons are compressed into the atomic nucleus and are absorbed by the protons. This process of electron absorption is turning all protons into neutrons. As a result of this, the compressed central part of the exploding star becomes compact as an atomic nucleus because is formed only of neutrons. This extremely compact central region remains after the explosion of the star as a Neutron star. If the star is substantially bigger, then instead of a Neutron star, in the center is forming a Black Hole. - This is a brief explanation of how mainstream science is explaining the Supernova explosion and the birth of the "Neutron Star". This explanation looks very good and logical. The problem is there, wherewith the advance of our astronomical instruments some facts and observations cannot fit in this explanation and is creating "Puzzles" and

"Anomalies". For example:

The modern powerful telescopes have discovered that even at 12.8by distance there are well mature galaxies with second-generation stars. To cover up this obvious catastrophe of BB theory, mainstream science declares, that all first-generation stars have been super-giants with a very short life. The problem is there were at the end of their life, the giant stars when collapsing suppose to turn directly into a black hole without any visible explosion. In this case, all of these first giant stars must turn into black holes and there no matter will be left for the formation of the Universe. (As usual, the facts which opposing the validity of their theories are just ignored!)

The next problem is that the official theory claiming that the iron and all heaviest elements are created by supernovas explosions. The problem is there, where there are not enough supernovas to explain the abundance of heavy elements in the Universe. The observed ratio between newborn stars to supernovas is about 400 million to one! - Which in no circumstance can explain the abundance of the past iron elements in the Universe. If we can just ignore this insignificant number of supernovae, then the question is - wherefrom the heavy elements in the Universe come from?

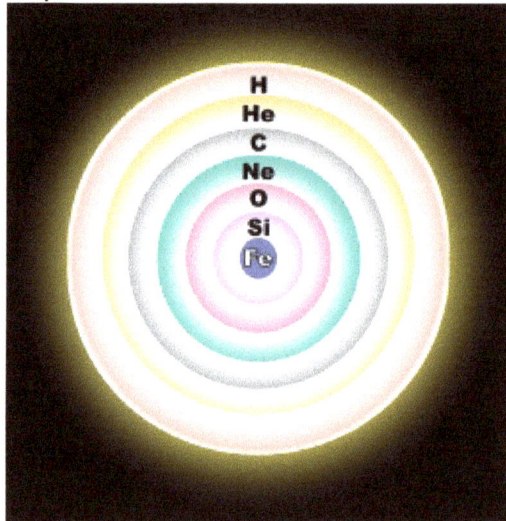

This is the sequence of how the elements
suppose to be produced in the Stars

We have to consider carefully all the available facts because any anomaly and puzzles in scientific theory are actual proof that this theory cannot be correct. Let see what the new findings of astronomy are and are they confirm or contradict the official explanation:

We have to start our consideration with the picture on a grand scale of the Universe:

On average, in the observable Universe, the production of new stars is about 400 million stars per day or 4,800 per second. The observable Supernovas are only about 300 per year! This is giving us a ratio between newborn stars and dying stars to be approximately 400 million to one! - This is a big discrepancy because the rate of dying stars and the rate of newly form stars must have a realistic and acceptable ratio. - This is the first fact, revealing that the official model for the formation and dead of stars is not correct.

Anyway, even when we are following the official model for the stars (where is many different classifications for the stars and different prediction for their life span). We will find in this model, that the average number of 0.1% of the existing stars will end their life as Supernova. Even if we assume that only 0.1% of the stars will end their life as Supernova and apply for this number on the number of the newborn stars, this is giving us the number that we should observe about 400,000 Supernovas per day! - The problem is there, where we are observing less than one Supernova per day! - This enormous discrepancy between the number of newborn stars and the dying stars as Supernovas is a revealing fact that <u>the Standard Model of Cosmology is 400,000 orders of magnitude out of reality</u>, and theory with such enormous anomaly between predictions and facts cannot be correct! Why there is such a big difference between the theory and observations and nobody is noticing that? - I believe that many astronomers know that, but they are afraid to speak out!

It is getting very interesting because for a long time we was believing that we know very well how the stars are born, how they are producing their energy and how they are dying, but suddenly, a few new finding is putting a big question mark on our assumed "knowledge"! We don't have to be afraid of fact, which is disproving our concepts because these facts have the potential to give us complete new prospects of understanding of the Universe.

To find what is going on, we will consider the property of the closest to us star - our Sun, and the two closest to us and the most studied Supernovas - 1987A and Crab Nebula, which is a remaining of recent Supernova.

Our Sun is an ordinary star and the Sun's structure, in general, will be valid for most of the stars in the Universe. In the previous article for the Sun, we have found that the temperature sequence and the specific gravity weight of the Sun revealed that the officially promoted model of the Sun substantially is differentiating from the facts and cannot be correct. The facts and observations are telling us that Sun's body is formed mainly of liquid with a solid center. This configuration is ruling out the promoted concept of star collapse because the solid and liquid materials are not collapsing. This is a significant finding because these facts are ruling out the credibility of the officially promoted concept that the Supernovas are collapsing stars. Also, the lower temperature of the Sun interior is ruling out the assumption for nuclear fusion. So... what really is going on?

We are running into the question: If Nuclear Fusion cannot occur, then how the star is producing its energy? Looks like we are in a situation, where the official model is not supported by the facts. To solve this puzzle we need to consider more facts and evidence to be able to shed light on it. - Fortunately, there is a load of good and reliable evidence and we will be able to solve the problems and puzzles of the Supernovas:

In 1987 in the Large Magellanic Cloud about 168,000 light-years away from us has exploded a Supernova. This is the closest to us observed Supernova and is one of the most studied.

Surprisingly, the source of this Supernova was a Blue Supergiant star named - "Sanduleak - 69 202." According to the official theory, blue stars don't suppose to explode. According to the official model, giant stars as Sanduleak - 69 202 in the last stage of their life just imploding and forming a black hole without any visible explosion. - These stars suppose to disappear quietly without a trace. The observed explosion of a giant blue star is a fact, which is disproving the fundamental principles of the officially accepted model for the physical processes in the Stars... But the surprises didn't stop there! By the official model, the giant stars after the explosion must form a black hole. The surprise comes from the fact that after thirty years of careful observations the astronomers didn't find any sign of a black hole or pulsar. The remaining material of the Supernova has formed an expanding ring with a nearly empty center. And again the question is why the exploding star forms a ring? We know that a nuclear explosion always is forming a sphere! The other question is - why hydrogen and helium are the dominant remaining materials of this Supernova? These facts again were declared to be a "Mystery" because the proposed model of Supernova cannot explain them. - (See the image below)

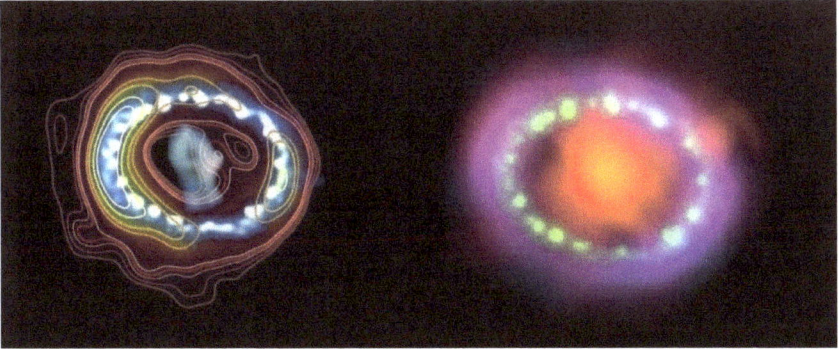

This is the image of 1987A with radio and optical telescope.
Notice the bright ring of balls, which even in the present are
centers of strong electromagnetic emission.

Unfortunately, the surprises for mainstream scientists didn't stop there. Two to three hours before the 1987A Supernova explosion the three underground neutrino detectors have detected extremely active - 10 seconds long neutrino burst. These neutrinos were coming from the Supernova and from a distance of only 168,000Ly. Despite this relatively short distance, the neutrinos come more than two hours earliest than the emitted light of the Supernova explosion! This is an undeniable fact, which is contradicting Einstein's assumption for the universal speed limit of light. To cover up this alarming fact, mainstream science comes with the explanation that the light of the Supernova explosion has been trapped inside in the ball of hot plasma for more than two hours. I am sorry, but this explanation is not credible, because the length of the Supernova explosion is about 10 seconds only. Further, we have incredibly rich experimental data for all types of nuclear explosions. - USSR and the USA have exploded above ground thousands of different types of nuclear bombs, and the most intensive light emission always is in the first a few milliseconds of every explosion. Mainstream science didn't explain also why the outer surface of the expanding incredible hot ball of plasma should not emit light for two hours? I would like to remind you again, that we shouldn't accept anomalies in our fundamental theories, because any scientific theory having such anomalies cannot be correct!
We have considered all range of the available evidence, there is not even one fact to support the officially promoted theory for the physical processes leading to Supernovas.
I have explained the reason for the Supernova explosion. This explanation is fully supported by our observations. Our observation is providing good confirmation of the electromagnetic origin of this catastrophic event. - Before

the explosion, around the star was a huge ring of dust and plasma. This ring around the star and the star is forming a capacitor on a grand scale. - (See the images above). This capacitor is receiving and storing the electromagnetic energy coming from the electromagnetic field of the galaxy. When the electrical potential comes to the extreme, a short circuit on a cosmic scale occurs and evaporates the unfortunate star. The other scenario is when the received energy exceeds the ability of the star to emit energy. The star starts getting hotter and hotter. When the temperature of the star reaches the necessary temperature for nuclear fusion, the star is exploding as one ordinary nuclear bomb. We should know that the magnetic field of the Universe on average is 60 times stronger than the Earth's magnetic field. The strength of the Galactic magnetic field is superior to the magnetic field of the Universe. The structures of the Universe are enormous, and these rotating in each other enormous magnetic fields are producing an enormous amount of electromagnetic energy and currents! Unfortunately, these powerful electromagnetic interactions are not considered by the Standard Model which is based only on the force of Gravity, which is a billion times weaker than electromagnetism.

Let see what the situation is with the other closest to us Supernova explosion and what it will tell us:

According to the "Standard Model," the stars are burning and transferring Hydrogen and Helium into heavier elements. The most revealing evidence which is ruling out the Nuclear Fusion processes in Stars is coming from the chemical composition of the Nebula remains of the Supernova explosion. The composition of the nebulas is a perfect representation of the original mater content of the exploded Stars. By study the composition of nebulas with good certainty we will be able to determine the original composition of the Stars. We have one such nebula in near proximity - Crab Nebula.

The Crab Nebula is a result of the recent and closest to the Earth Supernova explosion. Nearly a thousand years ago - in 1054y the Chinese astronomers have recorded the appearance of a new extremely bright star. The star has been visible even in the daytime with the naked eye. Modern astronomy knows that such a bright new star has to be a Supernova. This assumption has been confirmed when the astronomers pointed their telescopes toward the described area. There they found the remains of a Supernova in the form of a Nebula. This nebula has been named Crab Nebula. It is the most studied nebula and its chemical composition brings a huge surprise to modern astronomy.

Crab Nebula

The officially promoted model of stars is based on a nuclear fusion of hydrogen and helium content of the stars. They are assuming that when the star has exhausted nearly all hydrogen and helium and turning them into heavier elements, then the star exploding as Supernova. So... according to the official explanation in the remaining nebulas of the Supernova don't suppose to have Hydrogen and Helium!

After the explosion of the supernovas, there remain nebulas. The composition of these nebulas usually is about 90% hydrogen, 6% helium, and no more than 3%-4% other elements. This is contrary to the official claim. Such abundances of light materials in the nebulas we can have only when the Star is not fusing the lighter elements, but doing exactly the opposite - is disintegrating the heavy elements! - This composition is exactly the opposite of the official model and explanations. This fact is ruling out with 100% certainty the assumption that Sun and Stars have Nuclear Fusion. The remaining nebulas' chemical composition is telling us that more likely is that in the Sun and in the Stars working exactly the opposite process of electromagnetic fission. - Instead of fusing, the stars are disintegrating the elements into Hydrogen and Helium. To do this, the star needs a load of energy, which the star does not have. In this case become obvious that to be able to shine, the star needs constant input of external energy for billions of years. But the question is: can we find where this enormous energy is coming from? - Yes, Voyager Space crafts has detected a plasma firewall on the end of the Heliosphere with enormous electrical potential measured in tens of billions of volts! This sphere is enormous - 36 billion kilometers in diameter and is acting as a transformer on a colossal scale! We are lucky that our "transformer" is in a quiet region of the Galaxy and will not get overcharged, because if overcharged a short circuit can occur and evaporate our star.

I have explained above that the Universe has a very sophisticated energy/matter recycling mechanism. There are places where the light elements are fused together and are releasing heat and electromagnetic energy into the intergalactic medium. The plasma "network" of the Universe is ferrying and is transferring this energy between Galaxies and stars and is providing energy to the Stars. Under the influence of electromagnetic radiation, the Nebulas are producing a wide range of chemical substances including water and organic materials which are a necessary foundation for biological life.

From all the evidence which we have considered in the previous chapter for the Sun and the facts presented above, becomes obvious that not the Nuclear Fusion, but electromagnetic interactions are the prime source of stellar energy. The galactic electromagnetic field is incredibly powerful and the rotating star in such a powerful magnetic field inevitably will produce a lode of current and heat. That's why the hottest part of the Sun is its outer part - (the Sun's corona), with a temperature up to 2,000,000K.

The star's core is mainly liquid metallic hydrogen, in which the superconductivity is acting as a cooling substance of the star interior. This providing the stars with a very good radiation system, which acts as a cooling mechanism for the star's interior! That's why we are observing the temperature of only 4,000K in the sunspots holes. The liquid star interior cannot collapse and is not allowing the star to collapse and to explode as a supernova. The reason for the supernovas explosion is not a collapse! The reason is much, much different than the officially assumed version. Astronomers are observing that some stars are changing their brightness - getting dimmer or getting brighter. Recently, the famous star Betelgeuse has lost 30% of its brightness lately to gain it again. Some stars suddenly are completely disappearing. - For this reason, has been established the "VASCO Network" to trace these disappearing stars. They have found that about 150,000 stars in our galaxy have disappeared without any obvious reason. All these changes in brightness and star disappearing cannot be explained with Nuclear Fusion, but with the electromagnetic "malfunctions" of the electromagnetic energy supply between stars and galaxies. There is nothing mysterious in Solar and stellar energy exchange and interactions, when we are considering the facts with logic, with the laws of physics, and without preference and ignorance of facts.

There is no point for concern because our Sun is well balanced and stable. We are situated in a quiet region of the Galaxy and there is no reason our Sun not to continue shine for many billion years. The Earth will have a very long future if we do not make something stupid. When you are walking outside and enjoying a nice and sunny day, remember this article, raise your head up, and admire the Sun! - It is our light globe in the vast and unforgiving cold and darkness of the Universe!

## Neutron star, Pulsar, and Magnetar

These three types of stars are in the group of so-called "Neutron stars". All three of them belong to the same group; just they have slightly different properties.

The subject of Neutron stars is very interesting because we will find big puzzles and enormous anomalies not mentioned in other publications.

I believe that the explanation of Neutron Stars is a typical example of the adaptation of totally wrong physical concepts, which in combination with ignoring critical facts and suffering a lack of logic and common sense. - It is a hopeless attempt to explain the most powerful electromagnetic emission in Universe with the most unlikely elementary particle, which has no any electromagnetic property or ability to produce electromagnetic field – the neutrons! - How did such absurdity become an established scientific explanation?

The story of the discovery of Pulsars started with a big controversy. In 1967 a young student - Jocelyn Bell has notice some rhythmic cosmic radio emission. In the beginning her professor, Anthony Hewish was suspecting this to be a beacon of intelligent society and call them (Little Green Man). Shortly after the discovery of this rhythmic radio emission Fred Hoyle and Thomas Gold come with the proposal that this emission is caused by rapidly spinning "Neutron star".

The controversy of the discovery continues with awarding the Nobel Price not to Jocelyn Bell but to her supervisor – (Anthony Hewish and Martin Ryle) for her discovery of Pulsars. The unethical act of the two men who accept the prize and the glory for the young lady discovery has steered the astronomical societies. Even Fred Hoyle argued that Jocelyn Bell at least should be included in the prize.

Anyway, this is not an isolated incident with the wrong awarded Nobel Prize. Nobel Prize very often is used for political reasons or as a validating tool for some bizarre and weird theory. A good example is when the Nobel Peace Prize was awarded to Barak Obama, who started his presidency with two wars and ended with six. There is known a few Nobel Prizes for physics, aimed to prop up the impossible scenario of Big Bang theory - (as the superfast initial inflation), which is a brutal disregard of the law of Physics.

Another example is the award for the discovery of the Cosmic Microwave Background (CMB).

The discovery of CMB has been proclaimed to be a triumph of science and as undeniable proof of the correctness of the Big Bang Theory. Unfortunately, the fact is that CBM is proving exactly the opposite - that Big Bang never occurs because the CMB is coming from the opposite direction of the proposed location of Big Bang - (see the graphs at the end of the book).

Here is a short explanation of the official version of neutron stars, their property, and how they are supposed to forms:

At the end of the active life of a star, the internal radiation pressure of the star becomes weak, and this leads to sudden gravitational collapse of the star, which explodes as a supernova. In this violent event, the electrons and protons of the central core of the star are forced to fuse together and form neutrons. The enormous pressure of the explosion is pressing the neutrons tightly together to the extreme density of the atom nucleolus. As a result of this, the average diameter of the newly formed neutron star is about 12-20 km with 1 to 3 solar masses. A cubic inch of its material will weigh on earth about a billion tons. The pulsars are spinning incredibly fast! The spinning rate is measured by milliseconds! The fastest measured spin is 1122 rotations per second, which are 67,320 rotations per minute! The magnetic field of Pulsar and neutron stars is quadrillions of times stronger than the earth's magnetic field. The Magnetars' magnetic field is another 1,000 times stronger than this! The energy outburst from neutron stars is enormous by any standard! The Magnetar SGR 1806-20 is emitting in one second the energy our Sun producing in 1 million years! The source of this energy outburst supposes to be the kinetic energy of the spinning star and its interaction with the electromagnetic field (dynamo effect).- Fascinating story, isn't it? Just one "Small" problem! - Such energy output is exceeding greatly the well-known Mass-energy limit in Physics. In this case, the problem is coming from the understanding of the mass-energy equivalent. That means, that in a certain amount of matter is stores a certain amount of energy (Matter=energy) and you cannot take out of this matter more energy of what is stored there! -

Regardless of what kind of speculations the academia will present
Let see how credible is this "fascinating story," because we are given the most unlikely scenario for the event where the facts are showing something very, very different than the official story:

Will be good to start our consideration from the beginning - from the process which is creating the phenomena of the so-called Neutron Star. It is officially accepted that Neutron Stars are formed as a result of the Supernova explosion. After the violent explosion remains a cloud of hot Hydrogen, Helium, and one percent dust. The cloud - (the newly formed nebula) rapidly expanding and start cooling down. The heat and friction of the molecules are ionizing the gas and creating their own strong magnetic field. The Nebula is emitting all ranges of the electromagnetic spectrum - from gamma-ray to radio waves. And now the most interesting part of this story begins:

On a brief observation, the nebula looks like an ordinary cloud of dust and gas, but in reality, it is much more than that. The wide range of radiation is creating conditions for the production of a wide range of new elements and chemical substances. Neutral hydrogen atoms are very efficient in absorbing electromagnetic radiation. In contrast to them, the ionized hydrogen is pretty transparent. The different gas content of the nebula has a significant contribution to various chemical processes. The dust is absorbing the electromagnetic radiation on the lower part of the spectrum. The dust is heating the surrounding gas by emitting electrons due to the photoelectric effect. The dynamics of the nebula are very complicated and acts as a very sophisticated chemical space laboratory.

The ionization of the gas content of the Nebula is creating an extremely strong by-polar magnetic field and as a result of this in the central region start to appear the so-called "Neutron star" or Pulsar. These is not different stars - Neutron Stars and Pulsars are virtually the same, but they have a different appearance which is giving them different names. According to the official theory, the Neutron star, Pulsar, and Magnetar can be considered as a category of Neutron stars with slightly different magnetic properties. The difference between them is the strength of the magnetic field and the way how we can see them and their emission because the narrow emitted beam is visible only if the observer is exactly in the path of the beam.

Here is a list of some important facts about these stars:

- The first fundamental misassumption is that the stars are a hot ball of gas and at the end of their life they are collapsing. As we mentioned above, the specific gravity of the Sun is 1.4, and this is ruling out the credibility of this claim.

- The Neutron star is a dead star! - How can a dead star be able to produce continuously for millions of years billions of times more energy than an active star?
- How the electrically neutral particles of the dead star (neutrons) can produce a billion times stronger electromagnetic field than an active star which is formed of 99% electrically active plasma and liquid metal?
- The life of free neutrons not bonded in the atomic nucleus by the strong nuclear force is only 15 minutes! How could the neutrons can survive for millions of years in the neutron star and not disintegrate back into protons?
- The kinetic energy of the spinning neutron star supposes to be the "source" of its enormous energy! But the rotational kinetic energy of the collapsed star is less than its original kinetic energy before the collapse! – The rotational kinetic energy is not increasing by the collapse! There is not known any mechanism of continuous self-sustaining kinetic energy for millions of years! The law of physics is quite clear on this subject: - You cannot create energy! – In reality, academia is trying to convince us that the rotational kinetic energy of our Sun is trillions of times more potent than the potential nuclear energy of its enormous hydrogen content. - Are we should be so naïve to believe such weird claims?
- Sometimes pulsating emission of pulsars is speeding up; some time is slowing down, which is contrary to any logic! One of the official explanations is that the neutron star is absorbing matter. - I am sorry, but this is impossible! The centrifugal force of such fast-spinning objects will not allow this to happen! Another explanation insists that the iron crust of the neutron star is cracking and this is giving additional energy to the star? Such explanations are absolutely bizarre because they have never observed Neutron Star in detail. They don't know its composition and structure and such explanation is pure speculation. A simple explanation as a repeating discharge of a capacitor on a grand scale is a much more credible explanation but is unacceptable for them.

(See the image below)

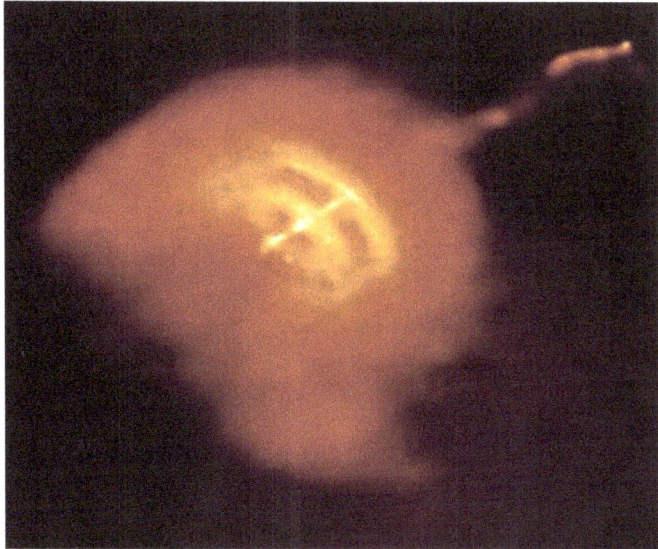

Vela Pulsar - here is visible the two brightest regions, which is acting as repeatedly discharging a capacitor of enormous scale

- The energy output of known neutron stars exceeds the mass-energy limit allowed by the law of physics by 1,000 times! Why should we believe an assumption which is known to be up to 100,000% out of spec?
- The average mass of a pulsar is about two Sun masses, but the average gravity of the pulsars is 6,000 times stronger than the Sun. A well-known fact is that gravity is related and is proportional to the mass of the object! How the theory of Neutron Star can explain such an enormous anomaly? An again, this model is "only" 300,000% out of spec in the case of gravity! Do we have to accept such enormous anomalies between explanations and facts? - Is this is what we call science?

The most powerful Gamma Ray burst (GRB) has been detected in 2019. This burst originated at 5 billons light-years away and is assumed that has been emitted by a Magnetar. In a few seconds only, this emission has released more energy, than our Sun is releasing in a period of 10 billion years, which is close to the predicted life span of the Sun! This fact is rising the serious question - how a dead star can release such an amount of energy and how the source of this energy can be just the kinetic energy of the spinning Star? - This is absurd!

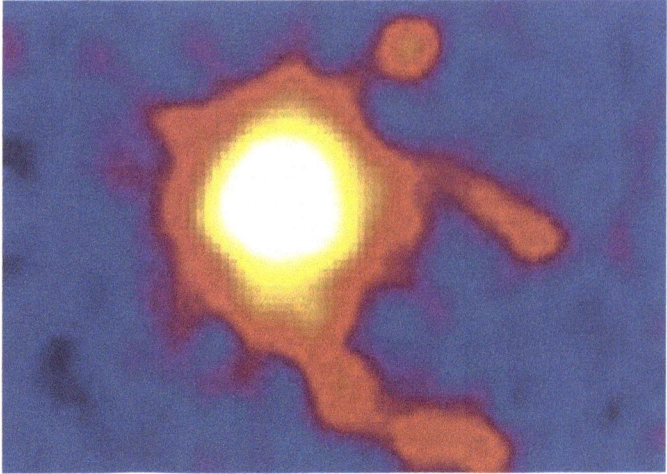

This is the most detailed photo of Pulsar which we have. It is obvious that such "detailed" photos cannot provide information for the structure of these objects as their "Iron crust"

Simple arithmetic can tell us that the rotational kinetic energy of the star is negligent to the enormous energy output of the Neutron stars. Further, the law of physics can tell us that if the radiation outburst of a neutron star is generated by the rotational kinetic energy this energy outburst will act as an effective rotational brake, and the neutron star will I lose its rotational energy in a matter of hours! The fact that the sequence of most pulsars' signals is more accurate than the atomic clock is undeniable proof that the source of energy is not the rotational kinetic energy. This fact is telling us that the energy source of the hypothetical 'Neutron Star' definitely is something else! - And this "something else" we have discussed and explained in the chapter of the recycling mechanism of the Universe - The Cosmic electromagnetism! To be easy to understand how absurd is the explanation that the energy source of the neutron stars is their rotational kinetic energy, I will give a simple example, which many ordinary people knows and will understand easy - When you start a petrol generator, the engine run free and happy until you start using electricity. The more appliances you connect, the more the engine starts struggling because the electricity consumption acts as an effective brake on the spinning generator! - How bizarre will be the claim, that when the generator runs out of fuel the kinetic rotational energy of the engine's fly-will will be able to run the generator for millions of years and the electricity output will be a billion times stronger, than when the generator is running on fuel? - The same scenario is valid for the proposed neutron star – it is obvious that the enormous energy outburst of Neutron Stars must have a continuous external energy supply or other ways it will act as an effective rotational brake and will stop the object spin in a matter of minutes!

Such absurd models and weird explanations as this we can have only when these explanations are based on incorrect fundamental physical principles. The lack of correct fundamental principles is the reason for us to be bombarded with such weird and illogical models and stupid explanations, where the most powerful electromagnetic fields in the Universe is produced by the most electromagnetically impotent substance - (the neutrons)!

Our modern sophisticated instruments are providing us with an enormous pool of data and facts, which cannot be fit anymore in the old "Standard Model" and its explanations. Will be good if we start asking our knowledge to be put on correct fundamental principles and the Law of Physics!

## QUASARS

Image of Quasar

Modern astronomy explaining that Quasars are the most energetic objects in the entire Universe. The Quasars are classified to be Supermassive Black Hole, surrounded by a disc of gas and dust with two powerful perpendicular jets. The falling in gas creates an enormous energy output. Quasars have incredible strong electromagnetic radiation <u>and can outshine thousands of galaxies for an extended period of millions of years</u>.

All quasars are assumed to be billions of years away from us because they have significant light red-shift. Some Quasars are in the center of some galaxy, some of them are in the disc of a Galaxy, and some of them exist separately. Mainstream science claim that all Quasars possessing substantial light

106

red-shift, which suppose to be an indication of their great distance from Earth and all of them are retreating away with great speed.

Some astronomers start expressing doubt about the claim for the extreme distance of all Quasars because the concept of the gravitational light red-shift is not a reliable method. Another scientist expressing concern, that the energy output of Quasars exceeding greatly, the possible energy which even Nuclear Fusion can provide. Among the various proposals are the speculations that Quasars are "Black Holes" or "White Holes".

Let see what actually we know about Quasars because the correct understanding of this subject is very important for our understanding of the Universe and the energy distribution there.

I will start with the story of the famous American astronomer Halton Arp.

Arp was studying the Quasars and was fascinated by the assumption that all the Quasars are very far from us, they are very massive, and are retreating at a very fast rate away from us. He starts suspecting that something with our understanding of distance measurement and the retreating speed of objects is incorrect because the Earth is insignificant on a universal scale and there is no reason for such behavior of the celestial objects. The assumption for the enormous distance of all quasars is based on their significant light redshift. Mainstream science using the light red-shift as an indication of distance, but this assumption is absolutely wrong! - The light is an electromagnetic wave and the magnetic field of the Universe gradually depleting its energy. In addition to this fact is well-known the effect of gravitational red-shift. That means the gravitational field of the Universe also will affect and deplete the energy of the traveling light. We also discussed that Space is a physical medium, and every medium will deplete the energy of the traveling waves.

In order to keep alive the nonsense of Big Bang Theory and to proclaim that the Universe is expanding, mainstream science deliberately is ignoring the effect of these three factors. One day Halton Arp has noticed something very strange - that one Quasar is fiscally attached to the near galaxy with a stream of dust and plasma. According to the measurement of the light red-shift emission, the two attached to each other objects have to be at a distance of two billion light-years distance? The fact, that the two objects are physically attached to each other is absolute proof that the distance measurement with light red-shift is wrong. Then he continues to study and has found many more similar examples. When Harp presented his findings to the US Astronomical Society, they immediately cancel his work contract; shut down his work, order him not to publish his results, and bans him from teaching Astronomy and having a job in the USA. Fortunately, Halton Arp has been admitted to the Max

Plank Institute in Germany and he manages to publish his work in the book "Seeing Red". This episode doesn't need comment and is a revealing fact of how carefully and brutally the establishment is guarding the wrong concepts in Physics and Astronomy.

The official explanation of Quasars, as usual, is based just on gravity and mass and is ignoring a range of vital evidence. The official explanation stating that Quasars are Black Holes feeding on gas and plasma. Their mass estimation is to be up to a billion times more massive than our Sun. Well...well... but there are some problems with this official explanation:

First, the existence of the Black Hole is not proven yet, and the law of Physics is not allowing the formation of such kind of singularity! - (Planck constant). Recently, mainstream science has announced that they have managed to take a photo of Black Hole. This looks like fantastic news! Looks that the existence of Black holes finally is proven once at per all! - Isn't it?... Unfortunately, this is the next false news. In the magazine - "New Scientists" from 13 April 2019 page 5 is an article about this "famous Photo". The article is revealing that the photo is a computer-generated image based on two pixels only! - Two pixels are just two dots! - Every educated person knows that it is impossible to make a correct image using just two dots! From two dots a computer can generate anything! - From white elephant to the mummy of Tutankhamen! I am sorry but such kind of "scientific proof" is not serving the truth!

The second miss-assumption is the assumed enormous mass of the Quasars. - The observations show that number of Quasars has been ejected from their host Galaxy. It is an impossible claim that such a "supermassive Black Hole" can be thrown out of a very loosely configuration of stars. There is no any known physical mechanism or force capable of ejecting such a "supermassive object" from a diffused cloud of stars! - This single fact is proving that Quasars cannot be such massive objects and their physical properties are shaped by strong electromagnetism!

The third misassumption (which we have discussed above) is the distance to the Quasars. Their light redshift is "mistaken" for fast velocity recession and huge distances, but the observations show that this is not correct.

The last fact is the enormous and continuous energy output of Quasars. According to the model, the energy of Quasars supposes to be the result of the gravitational squishing of the falling in mass. The Gravity force is the weakest force of all and to use gravity to explain the incredible strong energy output is a bizarre claim. Quasars energy output greatly exceeding the matter/energy limit allowed by the Law of Physics with a magnitude of a thousand times! - Only this fact is enough to give us the understanding that –

If the Law of Physics is telling us that the Quasars cannot generate that much energy, then we have to realize, that <u>the energy of Quasars is coming from outside</u> - in form of electromagnetic energy.

What actually are the physical properties of these mysterious objects? The Quasars simply are spinning accretion discs as "vortexes" of gas and plasma attracted by some powerful electromagnetic field and nucleus in the center. The fast-spinning plasma in this magnetic field is generating extreme heat and produces enormous energy output as two opposite jets. The incredibly strong magnetic field of the Quasars is responsible for the observable (electromagnetic light red-shift), which mistakenly is assumed as a great distance to this object. The incredibly strong electromagnetic field of the Quasars is producing also strong Gravity. This gravitational effect is giving the reason for the assumption that Quasars are as incredibly massive as Black Holes. It is obvious that the attempt to explain the Universe just with gravity and to ignore the role of electromagnetism is not working and is producing absurd models and unnecessary -"Puzzles". Honesty is the best policy toward knowledge! Only when we know the correct details of every object of the Universe, only then we can have its complete picture. - The picture in full color and beauty of our amazing Universe!

There is a way of how to see much further behind the "curtain" of CMB. Currently, radio astronomers are not allowed to look behind CMB, because the established dogma postulates, that this is the end of the Universe, and there is nothing behind! But there is a phenomenon, which will let us see a trillion light-years further behind CMB. - The streams of neutrinos, and the very powerful gamma-ray bursts. The incredible energy of these bursts is giving them the ability to travel on trillions of light-years distance before their energy is depleted. We have to learn how to read their information and to use them as "Standard Candles" for getting information on the distant structures of the Universe situated far behind the glow of CMB.


## GALAXIES, WHY THEY ARE NOT FLYING APART?

With the increased sophistication of our telescopes, we start seeing the celestial structures, which according to Big Bang Theory don't suppose to exist.   With the Hubble telescope, we manage to see mature second-generation galaxies at 13.4 billion years distance, which is just 400 million years after the proposed Big Bang. This is a huge problem for the "Big Boomers" because even for a galaxy of "first-generation" (the official

**classification is in reverse**) is necessary a few long times to form. To patch up their theory the "Big Boomers" come up with the idea of a mysterious and invisible substance - "Dark Matter", which gravity is acting as steroids for the formation of the celestial bodies. To justify their new "Invention" the Guys have declared that the galaxies are spinning too fast and their gravitational bond is no enough to provide the unity of the Galaxy and the galaxies should fly apart! This is a shocking statement because the guys were using the data of the luminosity of the galaxies to "determine" their mass and their gravitation bond. It is well known for all astronomers, that the luminous stars are representing just a small part of the star content of the galaxy because there is a big percent of dark invisible stars and the interstellar plasma filament is responsible for at least half of the total mass of the galaxy. With the emerging of the Hubble Telescope, <u>our estimating of the star content of the galaxies has increased thousand times</u>! For example, twenty years ago, astronomers believe that the Milky Way galaxy has 100 million stars, after ten years this number rise to 100 billion stars, and now the number grew to trillions. How with such a poor understanding of the mass content of Galaxies these guys were able to calculate "precisely" the gravitational bond of the Galaxies? Anyway, reason and honesty are not the first priority of these guys, and by ignoring and hiding facts they manage to "Establish" the public view that if there is not a "Dark Matter" the Galaxies must fly apart. In the previous chapter, we have considered how the strong electromagnetic field of the "Neutron Stars" is producing 3,000 times stronger gravity than the mass content of these stars. Obvious such revealing facts are ignored.

OK, they managed to "establish" their mysterious and undetectable substance, but the problems still remain! - When we looking at 13.4by away - where must be just a developing universe, space is full of mature galaxies and "second-generation" stars? - Why? The first stars must have only hydrogen and helium, only "second-generation" stars have heavy elements. The other problem is that we cannot see this "Dark Matter" and cannot detect it at all! - It contradicts all observational data, the law of physics, and common sense, but the smartest guys insist that this mysterious force is real and must be there. Do we blindly have to believe them? Are they preaching religion or science? Let see how they have "calculated" that the galaxies must fly apart? I will start the explanation with Keller's law for planetary movement. Kepler calculated and formulated his equation for the dependency of the planetary rotational speed to their distance to the Sun. The equation is simple and relatively accurate. The catch-22 is that, in the solar system, 99.9% of the mass of the system is located in the center. - The sun has 99.9% of all mass of the

Solar System, and the mass of the planets is only 0.1%. In this scenario, the mass of the rotating planets is insignificant, irrelevant, and even can be ignored. The center of rotation is in the center of the Sun. This is making Kepler's equation to be relevant to the Solar System only! What is the case with the rotation of the Galaxies? Can we apply to them the same formula that we apply to the Solar system? - Definitely not, because the mass of the galaxies is relatively evenly distributed in the disc of the galaxy. There is assumed to be a black hole in the center of the galaxy, but it is not more than an incredibly small fraction of the total mass of the galaxy. The evenly distributed stars in the disk of the galaxy are close to each other and form a gravitational and electromagnetic bond. This attractive bond acts like glue and forms one interconnected structure. The start and plasma filament of the galaxy is behaving like one rigid body. The gravity attraction is not purely in the center of this system how the case of the Solar system is. We cannot apply Kepler's law, designed for the Solar system where all the mass is in the center, to a structure where the mass is nearly evenly distributed in the disc, and where the parts have an electromagnetic and gravitation bond. It is simple, logical, and obvious! That's what we are observing! - Nearly all parts of the galaxy are rotating together because they are connected with strong internal electromagnetic and gravitation bonds. This is logical and obvious.

The rotation of galaxies has been accepted as normal a long time ago, and nobody has seen anything strange there until the smart guys intervened, and with the aim to justify the nonsense of the Big Bang Theory, they declared that this is a "big mystery." What happened, how come that madness has prevailed to such an extent? They have inventing mystery on top of mystery, and now the public has to provide them with further funds, big salaries, and expensive facilities to study their invented mysteries?

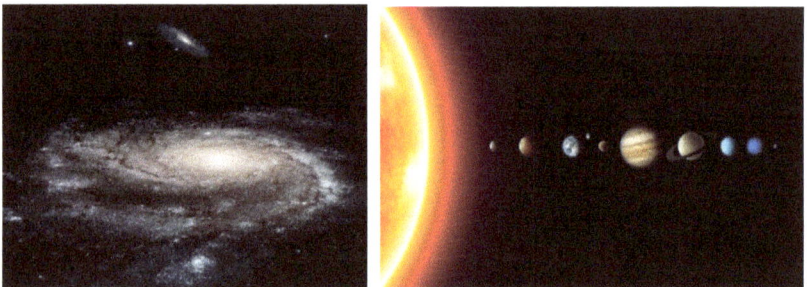

Compare the dense body of a galaxy and
the empty structure of the solar system

If somehow, in a public lecture, you ask some legitimate questions, they philosophically will tell you that this is a very complicated subject and is beyond your knowledge, your intelligence, and your ability to understand! - In real terms, they are telling you that you are stupid and you are allowed only to provide them with funds, big salaries, and to keep quiet.

This is the real story of how "they" have invented the mysterious Dark Matter, which is invisible, undetectable, not interacting with matter, has no known origin, has no known building particles, doesn't absorb heat nor radiates anything, does not clamp together to form solid objects, but provides only the "necessary gravity" for the smart guys to justify something more absurd and bizarre even than this substance – an absurd claim of a Religious Theory, which postulate, that this huge and enormous Universe come from "Nothing" and can be compressed in a dot smaller than an atom!

Well done! Such a manifestation of arrogance and ignorance must be awarded not with one, but with a bunch of Nobel Prizes and thousands of medals for dishonesty!

The tragedy of our time is that nonsense and arrogance have prevailed and become our everyday reality. Some people accept this as "normal" (or new normal) and don't even want to be bothered to hear or to accept the truth. We cannot continue in this way! - This is a dead-end for humanity!

We have to do something to put an end to this situation! It won't be easy because the biggest obstacle is embedded in our reluctant acceptance of nonsense and corruption. There is a way how this can be done very elegantly and without revolution, but we have to start with the rejection of nonsense and lies and to embrace and promote truth and knowledge.

In the beginning, I have promised you to make fair and complete considerations of all facts and all components of the Universe. It will be unexpected, but we have to realize, that biological life is an active part of the Universe, and we - humans are an active part of the biological life of the Universe. It will be fair to consider the most important principles responsible for the organization of the intelligent biological societies of the Universe including ours. I know, that It will be a very surprising subject to be included in a Theory of Physics, but without this part, my theory will be incomplete of the most important and valuable component of the Universe - Ethics, and Democracy.

Why do I believe, that Ethics and Democracy are a part of the Physical Organization of the Universe? I will try to explain to you the reason why I believe this in the following chapter.

## DEMOCRACY, WHAT IS IT, AND WHY HAS HUMANITY NEVER BEEN ABLE TO ACHIEVE STABLE AND LONG-LASTING DEMOCRACY?

Socrates - the first known victim of the idea of Democracy

There is one common dream of humanity - the dream to live in a good, stable, and long-lasting Democracy. For thousands of years, people continuously tried to find and select a good leader with an honest team who will work in their interest. Unfortunately, this still remains a dream which never had been achieved. - Why?

The tragedy of reality is that when the new leader sits on the throne, he very quickly forgetting his promises and starts enriching himself and his entourage. This scenario has been repeated over and over during the millenniums, and nothing ever has changed. - It is more than obvious that this approach of selecting a good and honest leader is not working, because the governing system is designed for the rich minority and the corruption always to succeed and a single person or leader is powerless to fight the wrongly design governing system. Let see first where Democracy is coming from and to find out what the meaning of Democracy really is:

At present, most people believe that they know what Democracy is, and they also believe that we are living in democratic countries. This belief cannot be further from the truth!

A popular belief is that Democracy is an invention of the ancient Greek philosophers. This belief also cannot be more distant from the truth.

Humanity was living in democratic social order much, much early than ancient Greeks. We have to understand and realize that everything we, the present

113

generation are, and everything we have achieved, is a result of the democratic social organization of the past civilizations! - Humanity has existed for many millions of years and faced many challenges and cataclysmic events. Humanity did not perish, because our distant ancestors manage to unite themselves to cooperate and help each other. They have been exchanging skills and knowledge on how to survive and prosper.

Archaeological evidence of the democratic social order of the human groups we are finding everywhere around Earth. - The ancient religious texts are informing us of the wonderful democratic society of Atlantis, which influence we are finding all around the World. The perfect megalithic structures, which are build long before the biblical flood 12,800years ago still standing on the four major continents and are the silent proof for this. The evidence suggests that in the Stone Age there hasn't been a personal possession or belonging of tools, food, or virtually anything! - In this time, humanity has been living in perfect social order and harmony, where everything belongs equal to each of them! Even in present days exists Ubuntu movement. - It is based on native South African social order of equality and harmony with nature.

In the period of the great geographic discoveries, we found many isolated tribes still in the stone age of development living in absolute harmony with nature and among themselves. We were unwise to bring them our modern technology and our social order, which didn't bring them happiness and prosperity! Our social order has done exactly the opposite - bring them only sickness, misery, disunity, corruption, and poverty!

In our rich archaeological evidence we are finding that in the late Stone Age and also in the Bronze Age, the supreme ruler of the tribes has been the assembly of the elders. The tribe has a military leader, but he hasn't been the decision-maker in times of peace!

The harmony and unity of human society start changing with the establishment of agriculture and the accumulation of wealth. Greed becomes more powerful than ethics and social values. Greed not only has destroyed equality, but greed also brings slavery and wars! The social and moral values, which have created the superior, intelligent, and passionate human race becomes a distant echo of the past. From this point, humanity has entered a new era and adapts new social order where ethics and morals have no place. So... what actually Democracy is? In short terms, Democracy is a peaceful and harmonious co-existence of different social groups. This social order can be achieved only when **everybody has equal rights and an equal voice in the decision-making process.**

The principle of Democracy formulated by the ancient Greek philosophers is - That Democracy is a rule of the people!

There always have been brave and intelligent people who are trying to promote and spread the principles of Democracy. It turns out that this is the most dangerous activity in the World. In the past and present these people are methodically and brutally eliminated by the Elite.

The first famous victim is Athenian philosopher Socrates – he was sentenced to drink poison because he was promoting Ethics and Democracy. After that the list of the victims becomes endless! - In the middle centuries governments created an institution or "industry" of extermination called "Inquisition." It was followed by the Nazi Gestapo, the communist Gulags, and with the present strategy of stealthy and secret extermination of unwanted people by giving them "heart attack" or cancer with the electromagnetic guns used by the secret services of all "Democratic" governments around the World. It is obvious that the price of knowledge of democracy is very high and is well guarded! That's why is worth learning the difference between democracy and our present social structure labeled as "Democracy"

And now, when we have formulated the principle and definition of Democracy we have to go back and start from the very beginning to be able to understand where really the principle of Democracy is coming from.

Will be a very interesting prospect, because I am planning to involve Physics and the structure of matter in this subject.

I know that this will be a big surprise for you if I tell you that the principle of democracy is embedded deep into the structure of matter and the structure of the Universe. - Yes, I know, this is a very bizarre, weird, and surprising statement, but it is absolutely correct and I will make an effort to explain to you why.

When we have considered the structure of the Universe, we have reached the conclusion that from the six components of the Universe, only consciousness is a creative phenomenon and it is the creative element behind the existence of the Universe. It becomes evident that if Consciousness is the dominant element in the Universe, then the principles of consciousness will be fundamental principles of the Universe!

So... where exactly these principles are hidden?

I have to start by explaining one fundamental fact. - Fact is that the Universe one way or another has been created at some point in the past. If we put aside the speculations for parallel Worlds, we have to assume that before the existence of our Universe nothing has existed! - Everything we see, everything we fill, everything we sense, has been created! - We are not creating any phenomena of the world around us! We just using and sensing what has been created! - Matter, the Law of Physics, Emotions, Ethical Principles...!

It is very important to know that the principles of Love, Curiosity, loyalty, Harmony, Ethics, Freedom, and Democracy are property of Consciousness and Consciousness has existed long before our existence and these principles are given to us! - We are not creating them! It is incredibly important for us to realize and understand where these principles are coming from.

We know that when matter and antimatter get in contact, they annihilate each other and are releasing just pure energy. This simple fact is telling us that in the matter is no any solid substance, and the matter is just a well-balanced configuration of opposing to each other electromagnetic forces - (And nothing else!) From Physics we have learned that the structure of matter - the elementary particles are just circular electromagnetic waves, oscillating around the center with opposite polarity.

(See the diagram on the left below).

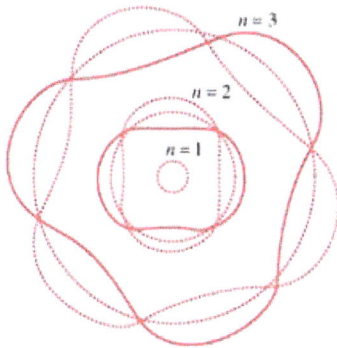

| The harmonic resonance of atom's waves | graphical representation of atom |

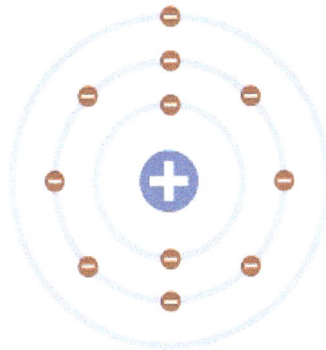

In order for this particle to be stable, the oscillating wave must have a certain frequency (resonance) to be in harmony with the center. If we have to describe with words what is the nature of elementary particles we have to say that it is a harmonic resonance of opposing forces! Or a shorter description will be the harmonic co-existence of opposing forces! This configuration we finding continuously in each particle, each atom, each molecule, and it continue up into the larger structures of the Universe as planetary systems and Galaxies! - It turns out that the definition of Democracy and definition of matter and the structures of the Universe are based on the same principle! - **Harmonic co-existence of opposing forces.** When we apply this principle to our society we can replace the term **forces** with **(groups)** - the basic formula remaining the same!

From the examples above, we have to understand that principle of harmony is the first and fundamental principle of the Universe in order to exist! - Without harmony, nothing can exist!

We, humans, manage to survive all mass extinctions and achieve everything which we have at present, by using and obeying exactly this fundamental principle in the time of the development of our civilization. Unfortunately, in the last few thousand years, we gradually departing from the principles of our success and as a result of this, we have managed to destroy most of the Earth's Eco-System. We are pushing the Earth's biological life in the condition of the fastest and worst mass extinction which ever been seen in the earth's history. We are altering the climate and making the future and survival of humanity absolutely uncertain!

The main reason why we have this misunderstanding of the principles of Democracy is the mass disinformation provided by the ruling Elite. Through mass media, television, history books, education, and philosophy the Elite manage to convince the people that they are living in democratic countries. To be able to understand the present situation, we have to start with understanding what Democracy is:

Democracy is a rule of the people! - Full stop! No more, no less! And no any other conditions!

That means that the will of the people must be the sole and supreme ruler of our society. Unfortunately, in the so-called "Modern Democracy" we have seen exactly the opposite, where a handful of privileged and immune politicians can produce any legislation and can impose their will over the people. - This is nothing by democracy! To get the full picture, we have to consider what actually the governing structures of the most advanced industrial nations are on Earth and to see if these governing structures are designed to serve Democracy?

Governing systems as the 'One-party system' of the communist countries cannot be labeled as democracy, because the one-party representatives are not elected, but they are selected from the top down. Unfortunately, the most populated and most industrial country in the World - China is one such example. The same system has Vietnam, North Korea, and Cuba. Their combined population is nearly 1/4 of the population of Earth.

Monarchy cannot be labeled to be a democracy also. To understand why I have to give you a brief explanation of how the Monarchy works:

The famous present Monarchy is the British Monarchy. The supreme ruler of the Country is the British Queen. The people voting there are advisory only! The Queen is the one, who is appointing or dismissing the Prime Minister, - (not the people). She is opening or dissolving the Parliament. Legally, the Parliament is ruling on behalf of the Queen (not on behalf of the people). Each bill or legislation of the Parliament first must be approved by the unelected members of the House of Lords. (The Lords are appointed also by Queen). Then each legislation must be signed by the Queen. Surprisingly (and unknown) for everybody, the British do not have a Constitution!

The Country is ruled by unwritten sets of rules which can be interpreted differently, depending on the circumstances.

When we are aware of the governing structure of the Parliamentary or Constitutional  Monarchy it becomes obvious that this system has nothing in common with any principle of Democracy.

Currently, 11 European countries still are Monarchy. In the World still, 43 Countries are Monarchy.

When you added to this list the communist country and the embedded social inequality of the "Caste" system of India, you will have a real picture of how "Democratic" is our World.

For example, the recent referendum in the UK for Brexit was declared to be an "Advisory" mechanism only. The result of the referendum was brought to the Parliament for approval. - Is this really an act of democracy and respect to the will of the people, or it is exactly the opposite?

Another example, the Judges and the Supreme Courts of most of the Western "Democracy" are appointed by the politicians. The Supreme Court is the highest level of power and is the supreme ruler of the Country. It is run by the most loyal to the politician's individuals. The Supreme Court has the power to overturn even the decisions of the Parliament! Is there nobody smart enough to figure out that to give such enormous power to unelected, by selected by the politician's individuals protected by total unaccountability is a supreme manifestation of tyranny and is representing the power of the minority over the majority? - In the election of George W Bush, the results of voting went to the Supreme Court. What kind of Justice is this, when the journalists were able to predict and named exactly how the Judges of the Supreme Court will give their verdict. - And exactly this had happened, the Judges appointed by the Republicans voted together, against the Judges appointed by the Democrats. - This is a farce, is not Democracy!

Here is another example - the UN Charter of human rights is giving rights of freedom of self-determination to nations. Despite this, in the recent Catalonian referendum for self-determination, the Spanish "Democracy" has declared that a referendum is a criminal act and the democratically elected people representatives were put in jail with the silent approval of all European Governments!

In Democracy, the people have to approve the ruling of the Parliament, not the other way around!

The British Empire has deliberately divided and separated too many nations following the rule - "Divide and Rule" - For example, most African nations are divided by wrong borders. The same is valid for the Middle East too. - Forty million Kurdish people have no right to have their own Country; The division of India and Pakistan drove out of home tens of millions. At present, the Palestinians are refugees in their own land! Recently Serbia was cut to pieces too! I just need to say that no one real democratic Government will create and tolerate such kind of injustice!

Similar events we are observing in countries which are not monarchy - What is the situation with the labeled as the "Biggest Democracy of the World" - USA? Since 1886 In front of New York standing toll - the famous Statue of Liberty, but what kind of liberty, and liberty to whom? - Women in America have no right to vote until 1920. Until the 1950es the voting rights were given only to property owners! African Americans have removed the last voting obstacles in 1964. In practice currently in the USA, the presidency and parliamentary positions are reserved for superrich individuals only!

The Aborigines of Australia were granted voting rights in 1962.

For the young people, this event sounds like had happened a long time ago, but these events had happened in my lifetime - in my own present time!

Recently in Russia, the President and Parliament have approved a new Constitution before the people have a chance to vote on it! It appears that the referendum there is also just an "advisory mechanism." and the will of the politicians is an unopposed ruler! We are witnessing how police armed with shields, buttons, and tear gas is beating and dispersing peaceful demonstrations in all these countries labeled as "Democratic". The governments silently are passing legislation, where if the people are not happy and wanted to protest, the people first must consult with the authority, must obtain permission from authority and only if is allowed only then the people can protest. - This is an absurd situation! - To ask permission to protest from the same people you need to protest! I have seen many peaceful demonstrators arrested and sentenced to jail terms just because they are not happy with what the authority is doing. Gradually, disapproval of the actions of the authority actions becomes a "legislated crime". Such a situation is the exact opposite of the principles of Democracy and is a clear sign that we are not living in a Democratic World.

Originally, the Constitution of the Country was the supreme rule and bastion of the rights and will of the people.

The Constitution must be an untouchable rule for the politicians. Unfortunately, even this last bastion of democracy has been hijacked and compromised.

I have explained above, that the principle of Harmony is a fundamental principle of the Universe for its existence and stability. We are children of the Universe and this principle is valid for us too. Without harmony, ethics, and tolerance, nothing can exist! - The Universe is based on opposing antagonistic forces, where the unifying force must be dominant, other ways everything will fly apart! On the same principle is build our society. - Our society is based on the same principle of opposing forces of greed and ethics. When ethics is dominant, greed becomes a constructive force and is driving innovations and progress forward, but when Greed becomes dominant, it is destroying everything which we have!

There is a hidden principle of corruption in our governing system. - It is the principle that the elected representatives have total immunity and unaccountability. This is giving them the freedom not to be bound to their pre-election promises and to do anything they want! The principle of immunity and unaccountability of the governments really mean that the legal system of governing is not designed to serve the people and politicians not to follow and obey their election promises.

There are countless philosophies and many different models for governing and social structures, but all of them are suffering from the same problem – they are just utopia! That means that they are just very nice and good wishes, but are neither practical nor workable because all of them are based on the same wrong principle – to elect an "honest" government without including there a legal system to control the government actions. - **This is the flaw of all utopias!**

The banks say that the opportunity makes the thief! - And this is the bitter reality of life because when loopholes and opportunities for corruption are embedded in all structures of the current systems we inevitably will get corruption! - To avoid this, we must implement a legal social system where politicians will not have an opportunity for political favoritism or wrongdoing. - This is the only way to make a good and fair governing system. These measures are not utopia, and are really simple, achievable, and effective!

If you need to learn them, you can find them in the book "Myths Lies Illusions and The Way Out."

My dear friend, I believe that my effort to give you one more realistic picture and correct understanding of our amazing Universe was an interesting and valuable experience and were worth every minute of your valuable time.

In the end, I just need to tell you one more thing - that biological life is the apex of the Universe. The intelligent mind is the apex of biological life, but the principles of Ethics are the apex of the Intelligent Mind!

## IS THERE INTELLIGENT LIFE IN THE UNIVERSE AND WHAT IS THE FUTURE OF HUMANITY?

I am trying to give you a complete understanding of the Universe. Without considering the existence of advanced intelligent life in the Universe, my job will be incomplete. Until the present day the question of - Are we alone in the Universe still is unanswered. In my book 'Myths Lies Illusions and The Way Out' I have included the article "The Secret Message from the Past". In this article, I have revealed that the ancient symbol of the hexagram, which is known as "The Star of David" originally, is not a Jewish symbol, but is coming from the deepest sources of human history with an explanation that this symbol is given to us from "The Gods" and this symbol is a "Cosmological Diagram."

The big surprise for me was when I realize, that this diagram really is a cosmological diagram, representing Space, Time, and Matter in the most simple, but incredibly sophisticated way. At that time I was the only person on Earth, who understand the six-dimensional configuration of our space and I was the only person capable to understand the meaning of this cosmological diagram. To be constructed such a diagram is necessary substantial knowledge superior to our present knowledge - this is a knowledge, which our ancestors cannot have! It cannot be an accident that such a sophisticated diagram coming together with the perfect explanation, that this is a gift to us from "The Gods" and is a symbol of the Universe, and the creation of life!

Every intelligent person will understand the enormous significance of this finding. This symbol is telling us that the Earth has been visited by a more advanced intelligent race! This symbol is telling us that we are not alone in the Universe. This symbol also is telling us, that there must be a very serious reason why the others don't want to communicate with us! The rest is left to us - we have to figure out why the others are refusing to be our friends! - The reason is obvious - If we are such vandals toward our own planet, if we are killing each other with and without reason, if we are deceiving ourselves with everything possible what will be the reason for the others to become our friends and exchange ideas with us? And what ideas we can give them?

My finding of the cosmological diagram is undeniable evidence for the existence of other intelligent life in the Universe and also for the correctness of the six-dimensional configuration of Space. We have also thousands of evidence, that in the past we have been visited. - Evidence, carefully guarded by the elite.

This is the cosmological diagram of Baalbek. It is the perfect graphical representation of the Universe. The hexagram is the six Space Dimensions, producing matter and life in the center.

All this is inserted in the circle of the Time Dimension.
Only incredible intelligence can produce such a simple and
beautiful diagram of the Universe

There is a countless eyewitness of alien interactions and alien flying crafts, but all of them could be subject to doubt. Even if I personally see aliens, talk to them, or touching them, I cannot claim with 100% certainty that my experience is real and is not an imagination or hallucination.
Unfortunately, all Governments and the establishment work together and suppressing and destroying any evidence or information for the existence or activity of intelligent visitors. They manage to hide and destroy most of the evidence on the face of the Earth...Fortunately; they haven't done this in the depth of the oceans!
I am happy to inform you that I have found undeniable direct evidence of the mutable presence and activity of another advanced civilization on Earth.  This evidence is on an enormous scale, they are deep in the ocean floor in a place where we still cannot reach and we do not have the technology to produce such evidence. This evidence is in the depth of up to 7km on the bottom of the oceans. They are trucks of enormous machines like a truck of tank or bulldozer, but the weight and size of these machines are on an unimaginable scale - the trucks are 10 to 20 kilometers wide, and the marks of the vanes are sunk deep in the ground a few hundred meters.  There is not any doubt that these trucks are not a result of human activity, because we cannot produce machines of such enormous proportion and because these trucks have been left in a few occasions from present back in the distant past up to 50 million years ago! These trucks are crossing the bottom of the oceans in straight lines, some of them are long many thousands of kilometers. There is visible that these trucks have been made at least on three different occasions where the freshest trucks are not older than 200,000years. There are a bit older trucks, which I am putting them to be made about 10 million years ago. There are also many, much older trucks, which are nearly completely covered by silt. We know, that the sediment in mid oceans is accumulating at a very slow rate. For sediment of a few hundred meters to accumulate is necessary millions of years, especially in the clear water in the middle of the oceans. This makes me believe, that the oldest trucks are made about 50 million years ago.
Everybody can see these trucks by themselves. - Just open Google Earth and zoom as much as possible deep in the ocean floor. You will see that the trucks are everywhere.
I have made a good high-definition recording of all oceans and send the video

to many safe locations in order not to disappear. I didn't want t alert the establishment of my finding, but because they are monitoring carefully my research and activity, I believe that they already know what I was found because when I try to see the trucks again, the resolution was substantially reduced. They cannot be able to wipe out this evidence, because there are endless records of the ocean floor in the private archives, in the archives of geological companies and institutions, scientists, and biologists.

When we obtain the knowledge that the extraterrestrials have visited us in the past on a regular basis, and probably they are still here, we need to make very serious considerations of our behavior and our future:

- The first thing is to realize that we are not alone in the Universe.
- The second conclusion is that interstellar travel not only is possible but is a reality!
- The encounter with many flying strange flying objects is a sign that the visitors have left on Earth probes to monitor our progress and activity (if they are not steel here).
- The next conclusion is that there is not an embedded time limit for the existence of intelligent society, because we have evidence that the same civilization has visited Earth at least three times in the last 50 million years.
- And the last and the most important conclusion is that if we do not destroy ourselves there is no other obstacle for humanity not to grow into an advanced ethical intelligent society and to join the others.

We have to understand the reason why the other civilizations don't want to talk to us! The reason is on an ethical basis - the principles of ethics and freedom are not allowing foreign intervention in the life of others.

The well-known fact of life is that if somebody doesn't want to help himself, the others cannot help him! (The same principle is valid in our case).

We have to realize, that we are monitored, and the others know much more for us than we know for ourselves. They are here! They are listening and analyzing every communication, every action, and understand our psychological profile better than we! They know that among us are many gifted, talented, ethical, and wonderful individuals. They know how good and rich is our culture, the beauty of poesy, music, bravery, and sacrifices of countless selfless peoples. These examples of pure intelligence and uncompromised ethics are the reason for the patience of the united intelligent society of the Universe. These gifted and brave individuals are our collective treasure! - They are the reason why the extraterrestrials are not wiping us out!

The clock of our existence is clicking its final hours. Unfortunately, there is a well-organized and well-coordinated effort of the elite to exterminate and silence our most valuable individuals. The elite are perfecting the methods to wipe out and suppress any attempt for promoting or organizing ethical movements.

We are observing that day by day the plan of the elite slowly and gradually becoming reality. Very soon, when the last ethical and brave man is silenced, there will be no more hope for humanity!

Do we still have a chance? I am not sure, because until each of us takes as his personal responsibility the duty to stand up and defend the principles of ethics, truth, and tolerance, we have no chance!

I have started this book with the image of Giordano Bruno, and I will finish the book with him again. I believe that now you understand why I am classifying him as the 'Greatest Man in Human history' - Because of his supreme intelligence and bravery! He has been tortured for eight years! He has a chance to save his life by denouncing his Theory of the Universe. He hasn't done this just to give us an example of how we have to guard the truth and knowledge!

I am not possessing his enormous intelligence and with my modest ability I have tried to finish his work and sincerely I hope that I have succeeded.

Giordano Bruno - the philosopher and astronomer.
He dies because he tries to give us the eternal principles
of the Universe - Ethics, and Knowledge

## FINDINGS AND PREDICTIONS OF THE THEORY OF EVERYTHING IN PHYSICS

- The Universe is steady - It is not expanding or contracting.
- The Universe is a closed physical system.
- The Universe possesses all the qualities and functions to recycle its structures and to be eternal.
- The age and size of the Universe are unknown.
- The material part of the Universe is based on a six-dimensional orientation of a concentrated form of well-balanced electromagnetic forces.
- There is only one fundamental force in the Universe - The Strong Nuclear Force! - All other assumed as existing different kinds of forces are just by-products of the fundamental Strong Nuclear Force!
- The Universe is constructed of six fundamental building blocks – three materials and three non-materials. The material components are Space, Time, and Matter. The non-materials are Consciousness, The Law of Physics, and the Universal Quantum Information.
- The origin of Space is the "Primordial Space Dimension" which is a uniform one-directional oriented medium.
- Our Space is formed of Six spatial Dimensions with specific space orientations.
- The six-dimensional configuration of Space is well balanced, steady, proportional, and uniform in all directions. The structure of Space cannot be influenced by any known to us force or energy fields – tat means that Space is not changing its shape and volume. Or in short – Space is constant and uniform. Space is not bending, expanding, or contracting.
- Space is an energy-dense physical medium that provides the mechanism of Energy, and propagation of waves and forces.
- Each Space dimension has its own specific space orientation, which is providing the separation of the elementary particles by their so-called "Spin".

- The Space energy field is providing the conditions of dense (energy pressure) for the mechanism of the Universal Principle of Physical Attraction and Repulsion.
- The Universal principle of Physical Attraction and Repulsion is based on an energy deficit or energy excess between bodies of matter and particles. It is the same attractive or repulsive principle for Electromagnetism, Gravity, and Strong Nuclear Force.
- Time is a separate dimension in which Space is situated. Time and Space are separated physical entities. They are not physically incorporated.
- Time dimension has only one-directional propagation and does not have other orientations of its energy field.
- Universal Consciousness is the prime reason for the existence of the Universe. - The Universe has all the physical property of a giant supercomputer, where Consciousness acts as an intelligent information processor and is a storage bank for the Law of Physics and Quantum Information.
- The initial cancelation of the opposing dynamic momentum of the Space Dimensions has release energy. This released energy is the source of all the matter of the Universe.
- Space dynamic momentum cancellation is the key behind the proportional amount of Matter to the Volume of Space in the Universe – Space and Matter are in strict proportional dependency.
- The related amount of matter to the volume of space is providing the condition of a closed physical system of the Universe, where the Dynamic energy-matter Exchange Mechanism is providing the eternal existence of the structures of the Universe. The age and size of the Universe currently are unknown.
- Every point in space can be a reference point, where the Time is "Now" and Space is "Here". - This means that every point of space has its specific Time and Space values.
- The Theory is explaining that from each reference point Space and Time dimensions are propagating outwards in all directions. This is explaining the so-called phenomena of "Time Relativity".

- Time is a dimension and its intensity, or its energy field also cannot be affected by the internal physical processes. - Time cannot be affected by Speed, Gravity, or Forces - how the 'Theory of Relativity' assumes, because Time is not physically incorporated with the elements of the Universe.
- The Theory is providing unification between Newtonian Physics and Quantum Mechanics - (the fine layers of space)The Theory providing explanations that the Law of Physics is the same for the Micro and Macro World.
- The Theory explaining why the Strong Nuclear Force is a "Confined energy" that's why has a very limited range.
- The atomic structure of Matter is based on Six Dimensional Space orientation, where not only the Electrons forming six stable Shells, but the atomic nucleus also is based on six levels of shells.
- In the atomic nucleus shell configuration, the neutrons and protons are forming separate shells. The shells of neutrons are separating the proton shells. The first shell (or the nuclear center) is formed of protons.
- The "Solid" state of particles is based on concentrated and well-balanced positive and negative charged forces.
- The so-called "Strong Nuclear Force" is based on the energy deficit, which appears as the "Missing Mass" of the Nucleus in elements up to Iron and as "Mass Excess" in the heavier elements in past Iron elements.
- The theory predicts that the Star formation and Stars energy output mainly is a result of electromagnetic interactions on a Galactic and Cosmic scale.
- The Theory predicts that not the Speed of Light, but the speed of Gravity is the ultimate speed limit in the Universe.
- The Theory predicts that the so-called "Weak Nuclear Force" does not exist.
- The Theory predicts that all particles have mass.
- The Theory predicts that Quarks are composite particles.
- The universal Principe of Attraction explaining the mechanism for the

light energy depletion of the so-called "Light redshift" observed when Light travel in Space.

- The Theory predicts that "Weak Force" does not exist.
- The Theory predicts that "Dark Matter" does not exist.
- The Theory predicts that "Dark Force" does not exist.
- The Theory predicts that a "Black Hole" does not exist.
- The Theory predicts that "Neutron Stars do not exist in the proposed form, and that they are center of extreme electromagnetic forces acting as repeated capacitor discharge.
- The Theory predicts that "Higgs Field and Higgs Boson" do not exist.
- The Theory predicts that "Gluon" and "Graviton" do not exist.
- The Theory predicts that monopole does not exist.
- The Theory is providing Reliable fundament for answers in all aspects of Physics and Astronomy.

**With the following images, I am providing evidence for the impossibility of the Big Bang scenario.**

According to the official model of (BB) we should observe the continuous and uninterrupted sequence of the "development" of every bright object! This claim should provide us the ability to see the uninterrupted development of every luminous object from its present position back in time to its first appearance! That means if we are observing an expanding universe, we will see every luminous object of the sky like a continuous bright line going away from us - from the present position and time, back in time and distance, towards the time of its first appearance!

The fact that we cannot see such luminous lines of the gradual development of every object is an undeniable fact that the expansion scenario of the Big Bang theory cannot be correct!

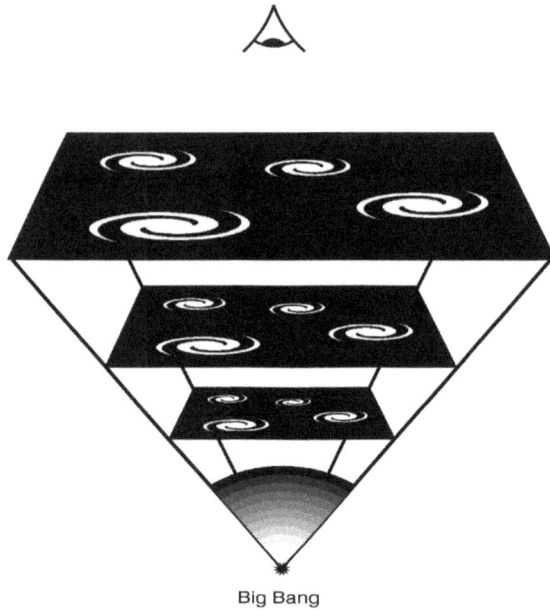

Big Bang

If the Big Bang theory is correct, the stars must get close and closer to each other toward the Big Bang location! But we are observing a homogeneous Universe in each direction, each distance, and each age!

Also, according to the Big Bang theory, we should be able to see the gradual development of the constellations in their previous position of time, (how is the image above), but instead of it, we are observing a totally different Universe on each stage and each distance! Such celestial configuration we can observe only if the Universe is not expanding!

On the diagrams below is an example of how NASA presents us with a false, misleading, and impossible model.

If our observation point is on the right side, how we have arrived there before the light (CMB) emitted by the Big Bang reaches this point? In this model also

131

in front of "us" are the stars, but behind is dark and empty space – how come? - Obviously, this is not representing the truth and cannot be correct! Further, they are claiming that CMB in the present is on 46by distance. They are not realizing that this means that CMB must be on 32billion years behind "BB" (in negative time)! Are they not intelligent enough to produce something sensible, or really they believe that we are so naive?

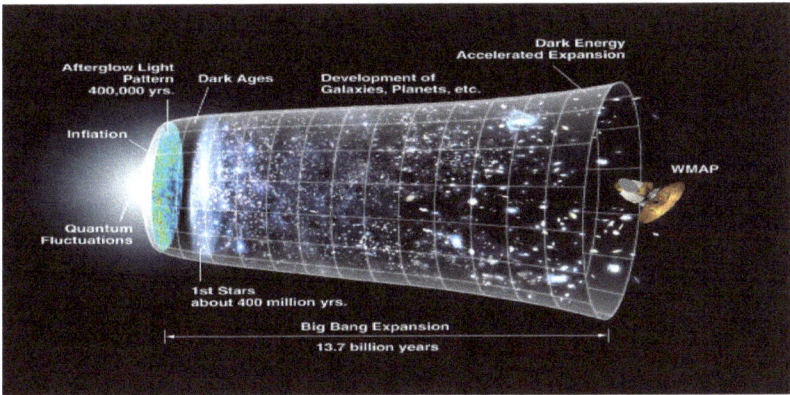

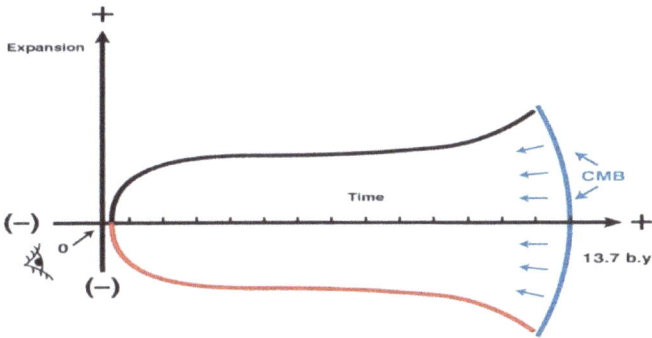

In this diagram is explained how NASA has constructed its misleading model: The black line in the upper part of the mathematical graph represents the hypothetical BB expansion. To add confusion, they have added the bottom part (the red line), which represents the nonsense of "negative expansion". Then they removed the graph arrows, added the stars, and switched our observation point from the left to the right side in order not to be figured out the correct origin and direction of the visible CMB, which is coming from the opposite direction of the hypothetical BB.
(See also the diagram below)

Here in this diagram is the correct position of our observation point and the correct direction where the (real), - the visible CMB is coming from. – Definitely not from the BIG BANG!

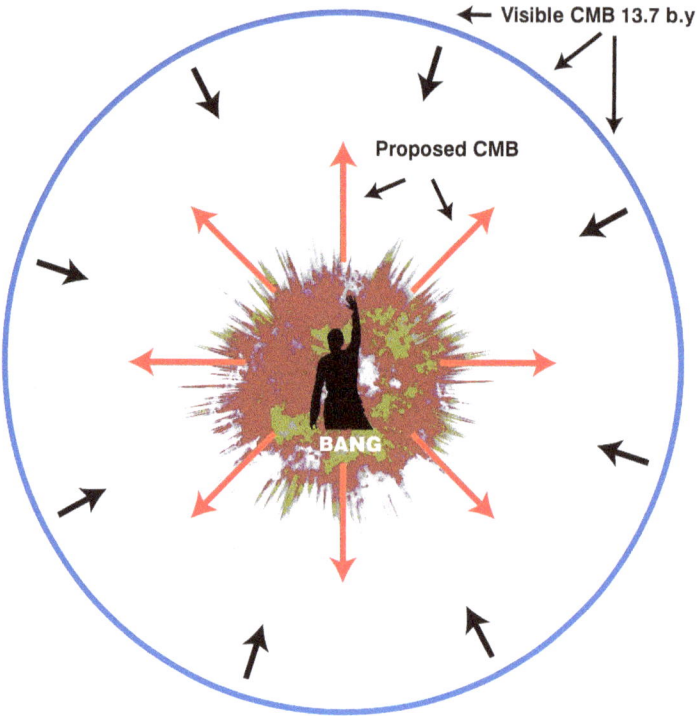

This is how the Big Bang diagram always should be presented - (as a cross-section)because the observable universe is a sphere, and we are in the center of it, but such configuration obviously will expose the nonsense of The Big Bang theory, because it is required at least 13.7 billion years of expansion time. In addition to this expansion, time needs to be added 13.7 billion traveling times of CMB emission to come back to us from the point of its emission (the dark blue circle) - The sum of those two figures of necessary times (expansion and light traveling back) is at least 27.4 billion years, which is double than the proposed
Big Bang theory scenario of 13.7 billion years! – Only this fact is ruling out the

validity of BB Theory!

Also in the diagram, above is visible the opposite directions of the proposed and the visible Cosmic Microwave Background Radiation (CMB) – (the red arrows) and the real, - the visible CMB, which is coming exactly from the opposite direction - (The black circle and black arrows) - So... how could CMB be emitted by "The Big Bang" if it is coming from the opposite direction of the proposed Big Bang?

We will never be able to see any CMB emitted by The Big Bang, simply, because of such light emission
is going away from us!

The visible CMB simply is the glow of the distant galaxies beyond our visible range, which is the proof
that our Universe is enormous, much, much older, and much bigger than the proposed 13.7by radius, and is undeniable proof that there never has been any Big Bang!

With the images below I am providing graphic explanations for the scientific incorrectness of the current officially accepted theories and evidence of atomic structures, which unmistakably indicating that the atomic structure of matter is based on six-dimensional space separation (orientation), where the revealing elements are with 3, 6 and 7 electron shells. Iron is the element where starting the elements with more than three nuclear and electron shells. - The mass deficit is ending with Iron, where the mass excess starts for the next elements. The perfect element is Carbon – is an element with **six** electrons, **six** protons, and **six** neutrons! Bismuth – it is the last stable element with **six** electron shells, and is where the elements with **seventh** electron shells are starting  - (Uranium) This is the reason for them to be unstable!

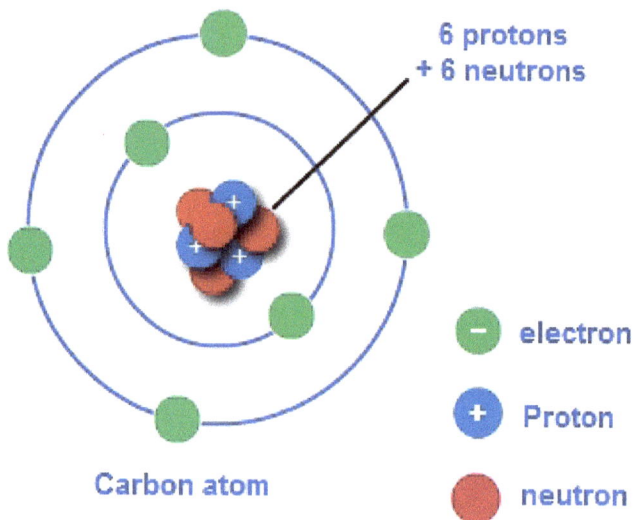

6 protons
+ 6 neutrons

− electron

+ Proton

neutron

Carbon atom

This cannot be an "accident" that the Carbon atom is constructed of SIX electrons, SIX protons, and SIX neutrons. - This is giving Carbon unique stability, versatility, and capability. This is the apex of the atomic structures because there are used all SIX space dimensions to accommodate and separate the atomic particles, and where the six atomic particles are not opposing each other. - This is the cleanest, sophisticated, and perfect atomic structure of all known elements. It is not an "accident" that Carbon is the most chemically potent element, and is the base of life, and all organic compounds. The mass deficit of the atomic nucleus is finish with Iron, were starting the fourth electron and nuclear particle shell. This is direct evidence for the energy cancellation origin of the attractive nuclear forces!

The two diagrams below are representing one of the best pieces of evidence for the SIX-dimensional space particle orientation, which provides the SIX different angular momentum (spin) of particles, and actually is separating them physically not to cancel and interact with each other in the boundary of the atom.

83: Bismuth          2,8,18,32,18,5

92: Uranium          2,8,18,32,21,9,2

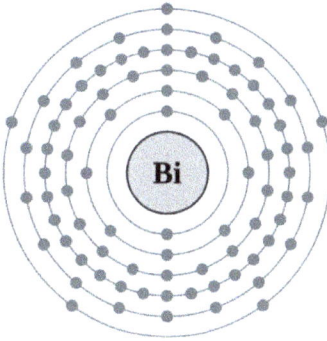

Bismuth – have **six** electron shells and is the last stable element

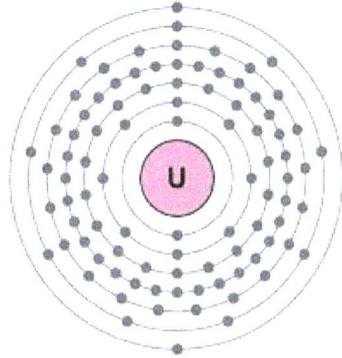

Uranium is the element with **seven** electron shells and this is the reason to be unstable

seven shells are unstable configurations because there are is only six space dimensional orientations! The seventh shell is going in direct "conflict" with one of the lower shells and the repulsion force becoming stronger than the attraction.

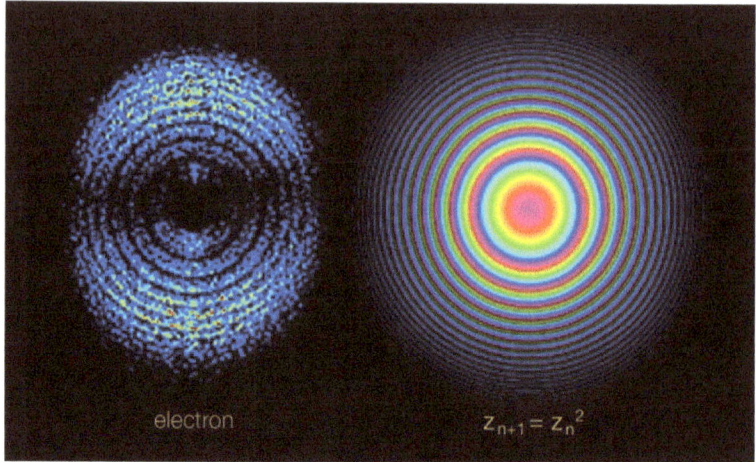

electron

$$z_{n+1} = z_n^2$$

This is a photo of an electron, which is direct proof of the concentric configuration of the standing waves around the center of space energy "vacuum," which is the concept explained above by the theory.

It's becoming obvious, that this is a common configuration for the microscopic world of matter –

(atomic and subatomic structures), where the combination of cleverly arranged fine layers of space dimensions and the centrally located point of cancellation of space energy is providing the condition of stable locked to each other opposing forces. This is the configuration, which represents the "solid" state of all particles.

(Curtsey of TOE) - Valentin Malinov 2017

Val Malinov is born in Bulgaria. His education includes: engineering, electrical, radio communications, building and construction, home sustainability, music, art, and chemists.
He is the inventor of a spherical combustion engine – 'Val Rotary Engine'
Val also holds a national bronze medal in white-water slalom.
He escaped from the communist regime of his native country with a 3 days solo journey with kayak across the Black sea.
Val is a member of the Astronomical Society of Victoria.
His special interests are: physics, quantum mechanics, astronomy, philosophy, classical music, art, and contact with nature.

This is the most revolutionary theory in Physics and Astronomy ever written.
The author insists that our understanding of the World doesn't have to be based on mathematical formulas, but on a deep understanding of the Physical processes of the Universe.
He reveals how the correct understanding of what is a single dimension leads to an understanding of the entire Universe.

www.ingramcontent.com/pod-product-compliance
Lightning Source LLC
Chambersburg PA
CBHW042117190326
41519CB00030B/7532